Why I Am A
Creationist
A Layman's Perspective

Chuck Nelson

Dedicated to: The Creator
Cover art/photography by Chuck Nelson

Library of Congress Control Number: 2020924814
ISBN-13:   Paperback: 978-1-64749-310-3
               ePub: 978-1-64749-311-0

Printed in the United States of America

GoToPublish LLC
1-888-337-1724
www.gotopublish.com
info@gotopublish.com

CONTENTS

INTRODUCTION

Surveys indicate, and my experience confirms, that many people believe in some form of creation or intelligent design as opposed to a purely naturalistic evolutionary model of origins. Often belief in creation or design is mixed with acceptance of some aspects of evolution and I have run into people who claim to be evolutionists but I find they also entertain some aspects of intelligent design or creation.

While naturalistic evolution appears to be the "official" model of origins in the United States and in much of the world, it is not always the "accepted" model of origins at the individual level. Many people are inclined to believe that creation or intelligent design has a significant role in the matter of origins. I think this belief is intuitive. When we see order and apparent design in the world around us, we intuitively think design requires a designer. If you were to see rocks on a hillside arranged to spell out "Free Lunch" would you think it was a chance arrangement or immediately "know" it was a product of intelligent design? Our life experiences tell us that it's not natural for rocks to arrange themselves to spell words.

Living organisms are immensely more complex than a few rocks spelling a couple words, yet we are asked to accept that life is an unintended natural chance arrangement of chemicals. We know chemicals can't fall in love, write sonnets, create artistic masterpieces, solve mathematical equations, tell jokes, build computers or found nations. We experientially and intuitively "know" that life is something more than chance and chemistry.

Our public education system and the popular media in the United States today promote naturalistic evolution almost exclusively to explain everything from the cosmos to people. Despite this, many people are not fully persuaded. Based on our life observations, experiences, and intuition, we know there's something more to life than chance and chemistry.

The creation vs evolution controversy generates a lot of heat and smoke. This is because basic worldviews are at the base of the issue. In this work I will be discussing both the science and the worldviews at play in the origins debate. I encourage the reader to get beyond the heat and smoke and understand the worldviews and philosophies at play. Take a close look at the data, the physics and fossils. What is the data really saying?

My purpose in this writing is to affirm and strengthen the intuitive belief in creation or intelligent design that so many of us retain. I have discovered that when you get beyond newspaper science and examine the facts for yourself, there is good cause to be a creationist. The facts are quite favorable to the creationist perspective and we are not alone; many scientists are firmly in the creationist or intelligent design camp. I hope to instill in others some of the excitement I have when I look at the science of origins through the eyes of a creationist.

In this work I will be able to present just some of the reasons I am a creationist. I have done a lot of reading and study on origins, creation, design and evolution and I will be sharing some of what I have learned. I will not be writing as a scientist or a theologian but as one layperson who has studied the issues for years. I will be writing as a layperson to other laypeople. I will not be trying to prove the existence of God; I will just point to some of the reasons it is a reasonable belief. I will be discussing some subjects and using some terms that may not be familiar to some of you. Don't worry about it, the concepts are easy to understand, and the unfamiliar words are just placeholders for understandable concepts. There will be no test at the end of the book so, no worries. Don't bounce off the unfamiliar words, just grasp the concepts.

If the reader already accepts creation or design, I hope this work affirms and supports your position and challenges you to further study. If the reader is a confirmed evolutionist, I hope to present new perspectives and information that you may not have considered. I will try to insert a little humor here and there, in part to keep things interesting, but in part because I can't help it.

I have tried to keep the chapters short so you can take small bites as you go. Some of the subject matter may be unfamiliar to you so short chapters should be easier to digest. I hope you find this subject matter as interesting and fascinating as I do.

I will end each chapter with some recommended resources for further study on the subject of that particular chapter. If the material in a chapter raises questions that aren't answered to your satisfaction, search the websites provided or books recommended for the answers you seek.

CHAPTER ONE
ORIGINS AND INTUITION

Many people find belief in creation or design to be intuitive. I have found this to be true even with people who have no particular religious affiliation or well-defined concept of God. It seems to make sense to them. At the same time, many elements of our culture reject creation and characterize it as a purely religious belief without scientific merit. People who believe in creation or intelligent design are often characterized as religious fanatics, anti-science, just plain ignorant or, at best, not intellectually informed. As a result, people with an intuitive belief trending toward creation or design often find themselves in somewhat of an intellectual bind. The idea of creation or design makes sense to them, but they aren't and don't want to be considered ignorant or anti-science. As a consequence, many people who harbor thoughts of creation or design in the origins debate tend to keep their thinking to themselves.

My purpose in this work is to validate the intuitive belief many have in creation and help them find their way through the philosophical, theological, and scientific matrix I will refer to as the creation-evolution debate. I will try to keep this discussion as non-technical as possible because we will be dealing with subject matter that most people don't encounter on a regular basis. I am not a scientist or a theologian but the concepts and issues in the origins debate are not beyond the reach or understanding of the average layman. I am writing as a layman to laypeople. Another way to say that is, "If I can understand it, anybody can."

I am a creationist. As far back as I can remember, I have always believed in creation even when my concept of a creator and creation was not clearly defined in my mind. Experiencing life and nature while growing up on a cattle ranch in northern California, the idea of creation seemed intuitive. Of course, the concept of creation requires a creator, so I believed in God even though my concept of God was sketchy. My family never went to church or seriously talked about God so my theology was limited to belief in some sort of God who would have been the creator.

I was given a Bible at an early age but when I looked into it, I quickly bounced off of the King James English and gave up on it. I wasn't given the Bible because of any particular religious commitment on the part of my parents. I think it was based on their thinking that a well-rounded individual should have some exposure to "religion" as part of their education. My point here is that my belief in a God and creation was not drilled into me by my parents, it was intuitive, and just made sense in the world I saw. Later in life, when I talked with my parents about religion, my father described himself as an agnostic and my mother was somewhat eclectic in her beliefs and dabbled in various religions including Christian Science, Unity and astrology.

As I moved through the public education system, my intuitive belief in God and creation was challenged but never fully shaken. I became a Christian through the influence of a girlfriend in high school who invited me to church. As I studied the Word of God, I came to know the God I had always known was there. Later, in college, my belief in creation was challenged and I compromised my thinking and decided God used evolution to create. I remember reading the following on page 21 of my biology textbook:

"Living creatures on earth are a direct product of the earth. There is now little doubt that living things owe their origin entirely to certain physical and chemical properties of the ancient earth. Nothing supernatural was involved – only time and natural physical and chemical laws operating within the peculiarly suitable earthly environment. Given such an environment, life probably had to happen. Put another way, once the earth had originated in its ancient form, with particular chemical and physical

properties, it was then virtually inevitable that life would later originate on it also." (Elements of Biology by Paul B. Weisz, McGraw-Hill, 1961)

I remember the stunned feeling I had on reading this. My science book was telling me that given the "natural" properties of matter, it was "inevitable" that life would originate on earth and that "nothing supernatural was involved." By that time, I had a strong faith in God and had received Christ as my Lord and Savior, but this appeared to be science challenging my intuitive and now theological belief in creation and God as the Creator. I wasn't able to give up my faith in God and I wasn't prepared to defy what I thought was "science" so I took the position that God used evolution to create. I was never comfortable with this position because I knew at some level that it clashed with both science and theology, but it allowed me to agree with both creationists and evolutionists at a superficial level. I was essentially whistling my way through a graveyard until I could deal with the matter in depth.

Many years beyond college, I began a personal study to see if I could resolve the creation vs evolution conflict that had never stopped troubling me. Although evolution was the only origins model presented through college, I never found the idea or evidence in support of it fully convincing and I was uncomfortable with my "God used evolution to create" position. Evolutionists claimed to know something about the origins and mechanics of life, but I heard nothing about how unconscious matter could become conscious and self-aware. I began to read books by scientists who were creationists presenting the scientific case for creation and against evolution. Wow! Where had this information been when I could have used it in college? I found myself being drawn into the sciences as they relate to life and origins and I continue to be fascinated by the subject matter. I found the case for creation by God to be powerfully supported by science. At the same time, I saw that the scientific case for naturalistic evolution is weak at best. The fact that certain chemicals can combine under certain conditions says nothing about how chemicals can form complex biological systems necessary for life, let alone think and reason. I was beginning to see that the text in my college biology book that had stunned me so many years before was not at all a scientific statement. It was a statement of a philosophical presupposition.

Like many people, I saw science as having great cultural authority and was not inclined to challenge science. However, at the same time I began to see where the discipline of science is frequently conflated with naturalistic, if not materialistic, philosophy. This is particularly true in academia and the popular media where naturalism is an assumed and often stated prerequisite for what some consider "true science". Gradually, I was beginning to distinguish between science, philosophy, and presumption. Presuppositions influence conclusions so it is important to see where the science ends and philosophy or presuppositions begin in the origins debate.

Over years of reading books by scientists who are creationists and studying the philosophical and theological issues, I evolved from an evolutionist, in the sense that I believed that God used evolution to create, to a Biblical creationist. I can't say I was a true evolutionist in my younger years because I maintained that the process of evolution was a mechanism designed and used by the Creator God. A true evolutionist in academia and the popular media today gives God no role in the origins debate. Remember, my college textbook cited only "natural" processes and specified that "nothing supernatural was involved."

I still enjoy reading books by scientists on the case for creation. Things are getting really exciting because the more science learns the stronger the case becomes for creation. More on this later.

I don't think my intuitive belief in God and creation as a youngster was particularly unique. Polls taken in 2007 revealed that some 60% or more of people in Canada and the United States believe in God and some form of creation. Given the fact that our educational institutions and the majority of the popular media are, for the most part, philosophically committed to the concept of naturalistic evolution, it seems amazing that so many still harbor thoughts of creation and design. I think that belief in God and creation is diminishing in our culture, but life experiences and intuition are hard to overcome. It's surprising that so many people entertain concepts of creation in the face of the antagonism to such positions in the media and public education system.

In this work I hope to affirm and support the reader who is inclined to accept concepts of creation or design in the origins debate. At the same time, I want the reader who is not particularly inclined to believe in the creation/design model to take a fresh look at the origins issue. To keep the non-religious reader comfortable, I will try to hold my theological position until the last chapter. I will not be entirely successful in restraining theological input, but I will try. Both creation and evolution have significant theological implications and can be found in world religions but I want the reader to look at and understand the philosophical and scientific issues at play in the origins debate without getting bogged down in "religion." Both the naturalistic evolution and creation/design models have strong "religious" implications so it's not possible to completely avoid them.

This work is a non-technical flyover of just some of the key issues in the origins debate, but my hope is that it will stimulate the reader to further study. There are many books written from the evolution, creation and design positions. This short work can't begin to address the issues in detail, but I hope it inspires the reader to begin a more detailed personal investigation.

FOR FURTHER STUDY

CREATION WEBSITES

www.icr.org
www.answersingenesis.org
www.trueorigin.org

INTELLIGENT DESIGN WEBSITES

www.arn.org
www.discovery.org/id/

Undeniable: How Biology Confirms Our Intuition That Life Is Designed, by Douglas Axe, published by Harper Collins.

CHAPTER TWO
DEFINING SOME TERMS

I will be using terms like creation, creationist, intelligent design, evolution, evolutionist, Darwinism and Darwinist in this discussion so I will try to identify what I mean by these terms in the context of this discussion. I will also define terms like naturalism or materialism that describe philosophical or worldview positions that come into play in the origins debate. These terms have somewhat of a semantic range or subtle differences in meaning but hopefully these definitions will help you understand my meaning in the context of this discussion.

I will begin my definitions with the term science.

SCIENCE: As defined by my Random House Webster's College Dictionary of 1992, science is: "(1) a branch of knowledge or study dealing with a body of facts or truths systematically arranged and showing the operation of general laws, (2) systematic knowledge of the physical or material world gained through observation and experimentation; (3) any of the branches of natural or physical science...."

When I searched Bing, I got this definition of science: "Science (from the Latin word *scientia*, meaning "knowledge") is a systematic enterprise that builds and organizes knowledge in the form of testable explanations and predictions about the universe."

As you can see, the definition of science is variable but focuses on knowledge of and study of the physical world.

Next, I will define the philosophical or worldview terms that are at play in the modern definition of science and at just about every level of the origins and evolution debate.

NATURALISM: Is a philosophical belief system that maintains that we live in a universe where all events, processes and reactions in the physical world are the product of natural laws. It's a somewhat mechanistic or reductionist belief system. Everything can be explained within the properties of matter, energy, and natural laws without invoking a supernatural (creator/designer) agent. Within this philosophical worldview, one can believe in a god and/or a spiritual domain if one likes but any such god is not the creator and not involved in the physical universe. In other words, God can exist if he likes but his presence and powers are limited to a spiritual domain and he can't monkey with the "real" physical world we live in.

MATERIALISM: Is a philosophy that is similar to naturalism in that it holds that everything that happens in the physical universe is a product of natural laws. However, materialism goes on to maintain that matter and energy and the natural laws that guide them are all that exists. Within materialism, there is no such thing as a god or spiritual domain. A more common name for materialism is atheism.

In today's public education system, many governmental institutions, and the popular media, the definition and practice of science has been conflated with the philosophical positions of naturalism/materialism. With these definitions in mind, you can see that when they are imposed as philosophical presuppositions on the discipline of science, the only possible origins model is naturalistic evolution. This is the belief system I have referred to as Darwinism. Any origins model that involves a creator or designer in any capacity is philosophically excluded from consideration, not by science but by definition. If the number 4 is equated with creation/design then 2 + 2 must be 5 because 4 isn't an acceptable answer. In the same way, when the scientific data related to origins are considered, only naturalistic evolutionary conclusions are acceptable within the constraints of Darwinism.

METHODOLOGICAL NATURALISM: This term has come into popular use with the advent of the creation/evolution origins debate. It's not in my 1992 dictionary but various definitions can be found on the internet and I'm reasonably sure it's included in science books in today's public education system. Here's a definition of methodological naturalism that I found on the internet: *"It is a ground rule that requires scientists to seek explanations in the world around us based upon what we can observe, test, replicate, and verify."* Philosophical naturalism is used as a "ground rule" today to exclude creation and design from the science of origins and evolution first because such concepts are not "naturalistic" and second because creation and design are said to be unobservable, untestable and unrepeatable. Never mind that the Big Bang and origin of life are not observable and repeatable. Creation and design in the origin and evolution of life can be studied and tested as well as the naturalistic origin and evolution of life. We are all looking at the same physics and fossils. The problem is any data pointing to creation or design is excluded from science because it doesn't conform to the "naturalistic" ground rule.

The term methodological naturalism is emphasized today primarily to accentuate the integration of naturalistic or materialistic philosophy with the discipline of science. It is often used to head off inclinations, interpretations or conclusions that point to creation or design. Methodological naturalism prevents the student, looking at the fossil record and seeing the sudden appearance of diverse organisms without transitional fossils, from postulating that it represents evidence of creation more than gradual naturalistic evolution. Concepts of creation or design are not allowed within the philosophical constraints of methodological naturalism. Consequently, scientific articles with conclusions or interpretations supporting creation/design are systematically excluded from most textbooks and scientific journals.

CREATION – CREATIONIST: Of course, a creationist is someone who believes in creation but there are many flavors of creationist. I am a *Biblical Creationist* in that I believe in creation by the God of the Bible and that the Bible provides a literal, if not detailed, scientific description of creation by God. I didn't start out as a Biblical Creationist, but I *evolved* into one by personal study of both science and Scripture. (See, I believe in some forms of evolution.) There are different flavors of Biblical Creationists, but I won't get bogged down with that here. There are also creationists who believe in creation by a deity or intelligence of some sort but not necessarily the God of the Bible.

I should point out that creation involves a "creator" and a supernatural act of creation that the so-called natural laws and processes we see in operation today could not accomplish. Creation, as such, does not reject the natural laws and processes we see in operation today, but it does not depend on them when it comes to origins. The physical world and the natural laws by which it operates today are the product of a supernatural creation. A creationist sees the so-called natural laws we know and study in science as conservational and operational not creative and innovational.

INTELLIGENT DESIGN: The Intelligent Design (ID) movement is largely found in academia where they use science, statistics, and logic to promote the evidence of design in nature, but typically don't try to identify the designer. They generally use science and logic to promote their position without quoting or using the Bible because they seek to demonstrate that design is a valid scientific position apart from religious convictions. Phillip Johnson, a law professor at UC Berkeley was an early leader in the ID movement and he referred to ID as a "wedge" effort, trying to make inroads in a rigid academic culture where naturalistic evolution had no use for a designer, let alone a creator. The ID movement includes people from a variety of religious persuasions, but it includes many Christians who are more comfortable or find it more strategic to use science and logic than Scripture to support their position. I have noticed some in the ID movement, while they support design, they don't seem to know what to do with the designer. They see obvious design in nature but appear uncomfortable with the "religious" implications of a designer and avoid trying to identify or describe him.

Despite the ID movement's efforts to present an objective scientific case for design, academia generally rejects the idea of design. This makes sense when you understand the prevailing naturalistic/materialistic religious or philosophical belief system in academia. For the most part the ID movement is considered to be a religious organization or movement promoting pseudoscience or anti-science and they are lumped with creationists.

EVOLUTION – DARWINISM: Evolution, as promoted and taught in the public education system and supported by the popular media is defined as a *natural* process by which life evolved on earth and diversified into the great variety we see in the fossil record and living today. Of course, the term evolution applies to the origin of the universe as well, but I will be focusing, for the most part, on the term as it applies to the origin and diversification of life on earth.

A key word in this definition, as previously noted, is the word "natural." The term evolution as accepted and taught in the public education system is an entirely natural process based on natural laws and the properties inherent in matter and energy. Remember how my college biology textbook specified it was a "natural" process and "nothing supernatural was involved." With the conflation of the discipline of science with the naturalistic/materialistic worldview, evolution as accepted in our culture and taught in our education system, has no room for a "designer" or "creator." The term Darwinism, as used in our culture is essentially interchangeable with the term evolution. When I use the term Darwinism or Darwinist, I am referring to the belief system or a person who accepts evolution with naturalistic or materialistic constraints.

You will recall that at one time I held the position that God used evolution to create. Many people today hold this position and will present themselves (as I once did) as a creationist or evolutionist depending on the conversational context. A person holding this position would not be accepted as a true evolutionist because evolution is a "natural" process, "nothing supernatural was involved," and there is no need for a creator or designer. From the perspective of Darwinism, the idea that God used evolution to create is retrofitting an unnecessary god to a natural process. To say that the origin and evolution of life is a natural process and God did it, strains logic and is not compatible with Darwinism where God has no role in the physical universe.

The person who wants to believe in both a creator God and naturalistic evolution (Darwinism) is employing a God who is orchestrating or guiding a naturalistic process which, by definition, cannot be considered naturalistic. Logic runs thin where one maintains it's a natural process and God did it. Darwinism, as taught in our public education system, tolerates no god who tinkers with the natural physical universe.

FOR FURTHER STUDY

CREATION WEBSITES

www.icr.org
www.answersingenesis.org
www.trueorigin.org

INTELLIGENT DESIGN WEBSITES

www.arn.org
www.discovery.org/id/

Intelligent Design; The Bridge Between Science and Theology by William A. Dembski, Published by InterVarsity Press

CHAPTER THREE

IS CREATION VS EVOLUTION
RELIGION VS SCIENCE?

In an ideal and perhaps idealistic world, the discipline of science would be a totally objective pursuit. However, in the real world, while we like to think of science as a totally objective discipline, scientists are not always objective. So, while pure science does not have an agenda, sometimes scientists do. Like everybody else, scientists have diverse beliefs, worldviews, political affiliations, and agendas. In addition to this, science is often practiced within or funded by public and private entities with their own objectives, agendas, and beliefs. This is not necessarily wrong or bad, but it can and does influence how and what science is practiced and what findings and interpretations are desired and accepted.

Scientific data never speaks for itself; it must be interpreted. However objective data may be, objectivity can be lost in the interpretation. Science tends to conform to cultural expectations. A contemporary example can be found in the way gender is viewed by today's physical and social sciences. It is said that if you torture data long enough it will eventually tell any story you want it to tell. We certainly see this in politics and sometimes we see it in what we call science.

We owe much to science and scientists and I don't mean to disparage either. My point is, we must evaluate and be slow to accept all statements, articles and even textbooks when they speak in the name of science. When necessary, or when in doubt, look beyond the statements and text to see if *all* the data supports the stated or published interpretation or if there is an alternative valid interpretation. To the extent possible, look at the political, cultural and worldview positions held by the entity sponsoring or funding the science.

When articles or books address the scientific case for origins or evolution, it is important to see were the science ends and philosophical beliefs begin. For example, my college science textbook said, "*There is now little doubt that living things owe their origin entirely to certain physical and chemical properties of the ancient earth. Nothing supernatural was involved – only time and natural physical and chemical laws operating within the peculiarly suitable earthly environment.*" The text went on to say that the origin of life was "*inevitable*" once the conditions were right. In the real world, life has NEVER been seen to originate spontaneously out of non-living matter and scientists using "intelligent design" have been unable to create life in the laboratory. The high confidence ("little doubt" "inevitable") in a naturalistic origin of life expressed in my textbook was not based on observable repeatable science; it was a statement of faith, a presupposition, methodological naturalism.

The conflict in the creation versus evolution debate is often portrayed as religion versus science but this is a false characterization of the issue. To a large extent, the public education system and popular media treats or portrays creation and intelligent design as non-scientific, pseudoscientific or anti-science religious positions. In order to understand this position, we need to understand the philosophical presuppositions in play.

Earlier I pointed out that the discipline of science in the public education system and popular media is conflated with and influenced by philosophical naturalism or materialism. These are terms we don't use in everyday conversation, but they are easy concepts to understand. I have defined these philosophies in chapter two so you can see the role they play in our thinking and how they influence science and culture.

Understanding these philosophies or worldviews and how they have been conflated with the discipline of science, one can see that the contention that "creation is religion and evolution is science" is an attempt to win the origins debate by definition before considering the science. Darwinists have been successful in capturing the terms of the origins debate by portraying it as religion vs science. In reality, both creation and evolution have significant religious/worldview implications but that doesn't preclude both positions from scientific evaluation. By "science" here I am referring to the actual data, the physics and fossils studied in the origins debate. I do not mean science constrained by naturalistic/materialistic philosophy.

When it comes to the study of origins, there are essentially only two options; some form of natural evolution or some form of creation. It is hardly true to the purposes of objective science (which means knowledge) to exclude one of the possibilities because it's philosophically unacceptable. I am aware that, theoretically, both natural and supernatural causal agents can be involved in both the creation and evolution models of origins but not according to the model of evolution presented in the public education system. The model of evolution presented in the public education system and generally supported by the popular media is limited by strict naturalistic presuppositions which preclude any consideration of design or creation. I have referred to the evolution model constrained by naturalistic/ materialistic philosophy as Darwinism. This form of Darwinism is the exclusive origins model taught in public education and generally supported by the popular media. Any attempt to interpret data outside of the naturalistic philosophical constraints is vigorously attacked, ridiculed, and rejected. This is why intelligent design is rejected as well as creation because both propose and cite evidence of a causal agent outside of purely natural processes.

I should point out here that Darwinists often claim they can't consider creation or design in the study of origins because they involve unobserved events and entities outside of science. This ignores the fact that origins science is historical science, it always deals with unobserved unrepeatable past events. We can't repeat and observe the origin of the universe or life, or the origin of the great diversity of life on earth, but nobody claims that such studies are outside the realm of science. We can look at the scientific data and come up with hypotheses or theories pertaining to origins. Science can and has always been able to distinguish design from random events and processes and there is no reason to exclude design as a possibility before the science is even considered. To do so is an attempt by Darwinists to impose their religion or belief system on others. Darwinists are saying, "My belief system is called science and your belief system is called religion." We don't have to buy their belief system when we study origins. Darwinists are attempting to win the origins debate by definition rather than objective science.

Regardless of how strong the data may be in support of a creation/design model of origins, any attempt to propose this within the public education system is rejected, ridiculed and shut down. It's said to be religion and outside the discipline of science when, in reality, it's not an invalid scientific consideration, it's a philosophically unacceptable consideration. Dr Scott Todd, an immunologist at Kansas State University illustrated this when he said, "*Even if all the data point to an intelligent designer, such an hypothesis is excluded from science because it is not naturalistic.*" Todd, S.C., correspondence to *Nature* 410(6752):423, 30 Sept. 1999. I am aware of similar statements made by Darwinists who are so invested in their religion of naturalism that they sacrifice logic on the alter of Darwinism for the salvation of their definition of science.

Any creation/design model or consideration challenges Darwinism and is, therefore, vigorously attacked. Even the slightest allusion to design or creation is attacked because if any element of design or creation is accepted as true, the naturalistic/materialistic underpinnings of Darwinism are falsified. In this sense, the origins debate is actually a battle of worldviews. Any model of origins that involves design or creation can accept and is not harmed by natural processes and laws at work. However, the philosophies of naturalism and materialism that underpin Darwinism are falsified if any element of design/creation is accepted as true. The heated attacks on creation and design are fueled by the fact that they collapse the naturalistic/materialistic worldview, i.e. Darwinism. Any claim of creation or design is an attack on the religion of Darwinism.

So, the creation vs evolution debate is not religion vs science, it's worldview vs worldview. One worldview maintains that a god or intelligence can play a role in origins and the other worldview maintains that only natural, unguided processes are involved. Rather than studying the evidence pertaining to origins and following it wherever it leads, the public education system has imposed its naturalistic worldview or religion as a presupposition before the evidence is examined. This is defended by claiming that philosophical naturalism is science.

Philosophical naturalism has become so conflated with the discipline of science that it's often difficult to separate science from philosophy when reading a science textbook or in the trickle-down science that finds its way into public consumption through newspapers, magazines and television programs. New discoveries and data are always interpreted within the constraints of the naturalistic worldview even when they can be valid evidence of creation and design. By the same token, evidence and data contrary to naturalism/materialism are often ignored or rationalized.

It's worth noting that a requirement of any valid scientific theory is falsifiability. This requirement holds that for a theory to be scientifically valid, there must be the potential for facts, data or circumstances to prove it wrong. Creation and design can be falsified if science can produce facts, data and probabilities that negate these theories. Darwinism, however, cannot be falsified in our culture because the alternative theory (creation/design) is philosophically unacceptable.

I have tried to make the point that the origins debate is largely a battle of worldviews. Scholars who advocate creation and design and Darwinists are all looking at the same data… the physics and fossils, if you will. Darwinists don't know something creationists don't know; they believe something creationists don't believe. Darwinists are interpreting the data with different philosophical presuppositions or worldviews which takes them, or I should say, constrains them, to entirely naturalistic interpretations and conclusions.

FOR FURTHER STUDY

CREATION WEBSITES

www.icr.org
www.answersingenesis.org
www.trueorigin.org

INTELLIGENT DESIGN WEBSITES

www.arn.org
www.discovery.org/id/

CHAPTER FOUR

IS PHILOSOPHICAL NATURALISM A NECESSARY COMPONENT OF SCIENCE?

Darwinists often insist that philosophical naturalism, or as they prefer, methodological naturalism, is a necessary component of science because science could not advance if scientists attributed everything they don't understand to a supernatural god or designer. Of course, this statement is true on the face of it, but it hardly reflects present or historical reality. Many scientists working today are people who believe in God and creation or intelligent design. Belief in a creator God or intelligent design does not and has not historically impeded science. Science began to make great headway when scientists believed a rational creator God created a rational universe that man could study and understand.

The insistence on conflating philosophical naturalism with science is just an attempt to win the worldview battle by definition. Darwinists are saying that you can look at all the data you want but you have to interpret it through our naturalistic worldview. This position has taken a strong hold in the public education system and the popular media. You could almost call philosophical naturalism the state religion.

Scientific papers that promote or even imply design rather than naturalism are systematically excluded from most textbooks and many scientific publications. I have exchanged emails with a biology professor whose tenure was terminated because she invited students to discuss intelligent design off campus. Highly qualified professors and administrators are often dismissed or removed from academic positions for supporting intelligent design or advocating the teaching of the weaknesses of Darwinism. For the most part, the naturalistic state religion is rigorously enforced in the public education system. Non-believers are not welcome. If you don't believe in and fully support the Darwinist state religion, you may have a difficult time getting or keeping employment in public education and some government positions. The book titled *Slaughter of the Dissidents* listed below under "For Further Study" gives many examples of discrimination against non-believers in the religion of Darwinism.

When science is constrained by philosophical naturalism, all scientific data must be interpreted within the naturalistic worldview and all conclusions are limited to natural causes. When science is limited or constrained this way, evidence supporting creation and design cannot be interpreted as such because that conclusion is philosophically unacceptable. How can science be considered objective when it sets out to study origins (or anything for that matter) while deciding in advance what it can't or won't find?

To make my point, I will give a humorous example to demonstrate how science suffers when findings and conclusions are limited by naturalistic presuppositions:

Let us say some scientists stumble upon Mount Rushmore having never seen or heard of it before. For the sake of the argument we will maintain that nobody has ever heard of it or seen it before, and they can find no written history of Mount Rushmore. Now they must decide how the faces of presidents came to appear in solid rock. Scientists philosophically open to a creation-design model would easily interpret the evidence to conclude that the Rushmore formation was the product of a creator or intelligent design. On the other hand, if scientists were to constrain their interpretations of the data to natural processes, they would have to propose circumstances whereby wind, rain, snow, freezing and thawing and time and chance carved the faces of presidents into a rock formation. They could study the data in detail but would never come to the right conclusion because the truth was philosophically unacceptable.

This is a silly example, but it illustrates how you can reach a wrong conclusion if you start with a wrong assumption. Darwinists consistently exclude any interpretation of the data that includes design or creation because it is philosophically unacceptable. Naturalism and materialism are philosophical positions apart from science, but they are taught in our public schools as a necessary component of science. Students are given the data and taught to view it through the constraints of philosophical naturalism and proceed to reason in a circle to find naturalistic evolution is the only scientific model of origins because they have excluded any other possibility.

I hope I have made the point that philosophical naturalism is not a necessary component of science. Natural laws and processes are a valid approach to science, but it is unreasonable to philosophically limit findings and conclusions. Belief in God, creation or intelligent design has not and does not impede science. In fact, many working scientists today are advocates of creation or intelligent design. Further, many of the findings, facts, and laws that science is built on today were discovered by scientists who believed in God and creation.

Darwinists often dismiss scientists that believe in creation as not being "real" scientists. By this they mean that they aren't doctrinal Darwinists. In reality, there are many working scientists today in many different disciplines, who believe in creation/design. Also bear in mind that in many positions a working scientist today who reveals his or her belief in creation or design or disagreement with Darwinism could face opposition and even termination. Look through the creation and intelligent design websites at the end of this chapter and you will find many books and articles written by contemporary working scientists who advocate creation/design. I own a book titled *In Six Days* (Master Books) with articles by fifty current working scientists in various disciplines who are creationists. Following is a partial list of Christian creation scientists from the Institute For Creation Research website at: https://www.icr.org/article/bible-believing-scientists-past/ This list is comprised of scientists of the past who have made major contributions to science demonstrating that, contrary to modern Darwinist contentions, belief in creation/design doesn't and hasn't impeded science.

SCIENTIFIC DISCIPLINES ESTABLISHED BY CREATIONIST SCIENTISTS

DISCIPLINE	SCIENTIST
ANTISEPTIC SURGERY	JOSEPH LISTER (1827-1912)
BACTERIOLOGY	LOUIS PASTEUR (1822-1895)
CALCULUS	ISAAC NEWTON (1642-1727)
CELESTIAL MECHANICS	JOHANN KEPLER (1571-1630)
CHEMISTRY	ROBERT BOYLE (1627-1691)
COMPARATIVE ANATOMY	GEORGES CUVIER (1769-1832)
COMPUTER SCIENCE	CHARLES BABBAGE (1792-1871)
DIMENSIONAL ANALYSIS	LORD RAYLEIGH (1842-1919)
DYNAMICS	ISAAC NEWTON (1642-1727)
ELECTRONICS	JOHN AMBROSE FLEMING (1849-1945)
ELECTRODYNAMICS	JAMES CLERK MAXWELL (1831-1879)
ELECTRO-MAGNETICS	MICHAEL FARADAY (1791-1867)
ENERGETICS	LORD KELVIN (1824-1907)
ENTOMOLOGY OF LIVING INSECTS	HENRI FABRE (1823-1915)
FIELD THEORY	MICHAEL FARADAY (1791-1867)
FLUID MECHANICS	GEORGE STOKES (1819-1903)
GALACTIC ASTRONOMY	WILLIAM HERSCHEL (1738-1822)
GAS DYNAMICS	ROBERT BOYLE (1627-1691)
GENETICS	GREGOR MENDEL (1822-1884)
GLACIAL GEOLOGY	LOUIS AGASSIZ (1807-1873)
GYNECOLOGY	JAMES SIMPSON (1811-1870)
HYDRAULICS	LEONARDO DA VINCI (1452-1519)
HYDROGRAPHY	MATTHEW MAURY (1806-1873)
HYDROSTATICS	BLAISE PASCAL (1623-1662)
ICHTHYOLOGY	LOUIS AGASSIZ (1807-1873)
ISOTOPIC CHEMISTRY	WILLIAM RAMSAY (1852-1916)
MODEL ANALYSIS	LORD RAYLEIGH (1842-1919)

NATURAL HISTORY	JOHN RAY (1627-1705)
NON-EUCLIDEAN GEOMETRY	BERNHARD RIEMANN (1826- 1866)
OCEANOGRAPHY	MATTHEW MAURY (1806-1873)
OPTICAL MINERALOGY	DAVID BREWSTER (1781-1868)
PALEONTOLOGY	JOHN WOODWARD (1665-1728)
PATHOLOGY	RUDOLPH VIRCHOW (1821-1902)
PHYSICAL ASTRONOMY	JOHANN KEPLER (1571-1630)
REVERSIBLE THERMODYNAMICS	JAMES JOULE (1818-1889)
STATISTICAL THERMODYNAMICS	JAMES CLERK MAXWELL (1831-1879)
STRATIGRAPHY	NICHOLAS STENO (1631-1686)
SYSTEMATIC BIOLOGY	CAROLUS LINNAEUS (1707-1778)
THERMODYNAMICS	LORD KELVIN (1824-1907)
THERMOKINETICS	HUMPHREY DAVY (1778-1829)
VERTEBRATE PALEONTOLOGY	GEORGES CUVIER (1769-1832)

NOTABLE INVENTIONS, DISCOVERIES OR DEVELOPMENTS BY CREATIONIST SCIENTISTS

CONTRIBUTION	SCIENTIST
ABSOLUTE TEMPERATURE SCALE	LORD KELVIN (1824-1907)
ACTUARIAL TABLES	CHARLES BABBAGE (1792-1871)
BAROMETER	BLAISE PASCAL (1623-1662)
BIOGENESIS LAW	LOUIS PASTEUR (1822-1895)
CALCULATING MACHINE	CHARLES BABBAGE (1792-1871)
CHLOROFORM	JAMES SIMPSON (1811-1870)
CLASSIFICATION SYSTEM	CAROLUS LINNAEUS (1707-1778)
DOUBLE STARS	WILLIAM HERSCHEL (1738-1822)
ELECTRIC GENERATOR	MICHAEL FARADAY (1791-1867)
ELECTRIC MOTOR	JOSEPH HENRY (1797-1878)
EPHEMERIS TABLES	JOHANN KEPLER (1571-1630)

FERMENTATION CONTROL	LOUIS PASTEUR (1822-1895)
GALVANOMETER	JOSEPH HENRY (1797-1878)
GLOBAL STAR CATALOG	JOHN HERSCHEL (1792-1871)
INERT GASES	WILLIAM RAMSAY (1852-1916)
KALEIDOSCOPE	DAVID BREWSTER (1781-1868)
LAW OF GRAVITY	ISAAC NEWTON (1642-1727)
MINE SAFETY LAMP	HUMPHREY DAVY (1778-1829)
PASTEURIZATION	LOUIS PASTEUR (1822-1895)
REFLECTING TELESCOPE	ISAAC NEWTON (1642-1727)
SCIENTIFIC METHOD	FRANCIS BACON (1561-1626)
SELF-INDUCTION	JOSEPH HENRY (1797-1878)
TELEGRAPH	SAMUEL F.B. MORSE (1791-1872)
THERMIONIC VALVE	AMBROSE FLEMING (1849-1945)
TRANS-ATLANTIC CABLE	LORD KELVIN (1824-1907)
VACCINATION & IMMUNIZATION	LOUIS PASTEUR (1822-1895)

How valid is the Darwinist's contention that nobody who believes in creation or design is a "real" scientist? Much of science today is built on the shoulders of these creation believing scientists. Belief in a creator God has not and does not hinder science.

FOR FURTHER STUDY

CREATION WEBSITES

www.icr.org
www.answersingenesis.org
www.trueorigin.org

INTELLIGENT DESIGN WEBSITES

www.arn.org
www.discovery.org/id/

Slaughter of the Dissidents subtitled *The Shocking Truth About Killing The Careers Of Darwin Doubters* by Dr. Jerry Bergman published by

Leafcutter Press. The book is volume I of a series on the subject and it outlines many cases where teachers and professors are dismissed or attacked for failing to fully subscribe to the doctrines of Darwinism.

Men of Science Men of God, by Dr. Henry M. Morris, Published by New Leaf Press.

CHAPTER FIVE
IS THERE REAL EVIDENCE SUPPORTING CREATION OR DESIGN?

There is significant evidence in support of creation and design in the world around us. Some of that evidence will be revealed in following chapters. Earlier I pointed out that the idea of creation was intuitive to me and it is intuitive to many people because it makes sense in the world we see and live in.

Given a choice of believing in a creator God or believing that a distant big bang exploded the entire mass and energy of the universe into existence out of a sub-atomic particle resulting in an expanding cloud of predominantly hydrogen gas which eventually evolved into people, I'll go with my Creator. I once heard Dr. Duane Gish, a creation scientist with the *Institute For Creation Science*, with tongue in cheek describe hydrogen as a colorless, odorless gas that over time becomes people. I believe it was G. K. Chesterton who noted the irony in the fact that some people reject a God who created everything out of nothing but believe nothing created everything.

In the world around us, we don't see unguided natural processes contributing to organized complexity. We don't see chemicals forming living organisms. We see that, over time, things tend to run down, wear out, deteriorate, break and die. Things don't tend to self-organize into functional complexity. The direction or trend over time is in the opposite direction… toward disintegration. We will discuss this trend later as the second law of thermodynamics, but perhaps this trend that we readily see around us is why creation or design is more intuitive than incredible functional complexity and organization resulting from a big bang explosion.

In our familiar world, we tend to think in terms of cause and effect. When we see something happen, we often want to know, "What caused that? In fact, there is a law of cause and effect that states, "Every cause has an effect, and every effect becomes the cause of something else." The relation between cause and effect is readily observable in our world so it is almost intuitive to think of a first cause to set the cause and effect chain into motion. In principle, an effect cannot exceed its cause quantitatively or qualitatively, so a creator God seems a more adequate first cause for the universe and life within it than a big bang explosion out of nothing.

This idea is described at www.icr.org/first-law as follows: "In simple terms, every cause must be at least as great as the effect that it produces—and will, in reality, produce an effect that is less than the cause. That is, any effect must have a greater cause.

When this universal law is traced backwards, one is faced again with the possibility that there is an ongoing chain of ever-decreasing effects, resulting from an infinite chain of non-primary ever-increasing causes. However, what appears more probable is the existence of an uncaused Source, an omnipotent, omniscient, eternal, and Primary, First Cause."

My point here is that there is good reason for people to have an intuitive belief in creation or design. It's consistent with our observations and makes sense in the world we live in. However, Darwinists often accuse people who believe in creation or design of being ignorant or uninformed. Sometimes we are accused of not believing in science. In reality we believe in science but not by their definition of it. I have seen where Darwinists have equated not believing in naturalistic evolution with not believing in gravity. It escapes them that Isaac Newton, a creationist, discovered the law of gravity. And, of course, gravity is observable, measurable and repeatable while the evolution of life isn't. You may have heard Darwinists claim that creation isn't science because it isn't observable and repeatable. If science is limited to what is observable and repeatable Darwinists have defined themselves out of science.

To this point I have described why I and many others intuitively believe in creation or design based on our observations and experiences in the world around us. In the following chapters, I will describe some of the evidence for creation/design from a more scientific perspective. These are just a few examples of the evidence from science that helped me "evolve" into a Biblical Creationist. I don't expect every reader to accept my Biblical perspective of creation but I do want to validate the clear evidence of creation or, if you prefer, design, that we can see in the world and in the various disciplines of science.

I can't begin to cover all of the scientific evidence in support of creation/design, but I will discuss some that were significant to me in my journey. I will try to avoid being too technical in my discussions as I want this information to be interesting and digestible by people who may not have examined some of these issues before. I will consider this work a success if it challenges or inspires others to further research.

FOR FURTHER STUDY

CREATION WEBSITES

www.icr.org
www.answersingenesis.org
www.trueorigin.org

INTELLIGENT DESIGN WEBSITES

www.arn.org
www.discovery.org/id/

CHAPTER SIX
THE UNIVERSE HAD A BEGINNING

For me, the most interesting and convincing evidence for creation is found in biology. I will discuss evidence from biology later, but I should set the stage with some other interesting observations.

In the last century there were two popular models of cosmology. Each model attempted to compensate for problems in the other. For a number of years, it was widely believed that the universe has always and will always exist. This was called the *Steady State Theory of Cosmology*. In this theory, the universe was observed to be expanding but it was believed that matter was continually coming into existence and forming stars and galaxies, so the universe retained its density and essentially had no beginning or end.

The steady state theory had a number of scientific problems and the big bang model is the dominant view of cosmology today. Today's big bang theory bears little resemblance to the original big bang theory. This isn't a bad thing as science must evolve with knowledge, but the big bang still fails to explain many observed phenomena. It's not a perfect model but it does employ methodological naturalism and avoids the God "problem." The big bang idea is constantly tweaked to try to find ways around conflicting data. Today's big bang theory holds that the entire universe burst forth from a singularity…a sub-atomic particle that contained the entire mass, energy, and space of the universe. This theory proposes that the universe burst forth as a quantum fluctuation of the vacuum of empty space. The concept of inflation – a faster than light expansion of the universe during the early stages of the big bang - has been added to try to explain contradictions with real world observations.

The idea that the entire mass, energy and space of the universe suddenly exploded into existence out of a sub-atomic particle in empty space is interesting. I don't know how often this happens, but wouldn't it be irritating if it happened in your nose one day?

Some Darwinists were concerned about the idea that the universe had a beginning because that implied a creation point and an opportunity to insert a creator. To keep the origin of the universe within the constraints of methodological naturalism an exploding sub-atomic particle containing the total mass and energy of the universe was inserted in place of God. In my opinion this seems like replacing a Creator God with a tooth fairy who pulled a universe out of a hat. In terms of the law of cause and effect, is a quantum fluctuation out of nothing a more reasonable first cause for you and the universe than a creator? The origin of the universe is not observable or repeatable, so the science runs thin and gets highly speculative and theoretical.

Apparently, as far as science constrained by naturalism is concerned, the universe is just one of those things that happens from time to time. Speaking of time, time as we know it only came into existence with the creation of the universe. We tend to think of time as we experience it, moving steadily in the present from the past into the future. However, time is a product of the physical universe and influenced by such things as mass and velocity. Our limited view of time is a consequence of our perspective in time and space. Time is relative but we operate well within our perspective of time so don't throw away your watch…. yet. We will talk more about the Big Bang in the next chapter and I encourage the reader with questions to search the websites below to learn more about the problems with the Big Bang model of origins.

As far as science can tell, the universe had a beginning. It hasn't always existed, and it is expanding from a big bang beginning. If we follow the timeline from today back past the big bang, we arrive at nothing! Everything out of nothing, there is no God to see here, move along.

FOR FURTHER STUDY

CREATION WEBSITES

www.icr.org
www.answersingenesis.org
www.trueorigin.org

INTELLIGENT DESIGN WEBSITES

www.arn.org
www.discovery.org/id/

CHAPTER SEVEN
A BIG BANG OR A BIG GOD?

I know I was going to try to restrain my theological positions until the last chapter, but I couldn't pass up the opportunity to use this chapter title as we discuss the Big Bang. As mentioned earlier, the Big Bang theory is currently the dominant model for the origin of the universe and us inhabitants. Apparently, universes are something that happens from time to time when there is a quantum fluctuation out of nothing, and we are darned lucky it happened. Where would we be without our universe?

A detailed discussion of the Big Bang theory would be very technical and beyond the scope of this work. My objective here is to validate the intuitive distrust that many people have of a theory that holds that everything exploded into existence out of nothing. The Big Bang is touted in public schools and the media as the scientific explanation for the origin of the universe. In fact, it is a cobbled theoretical construct with many weaknesses. Given naturalistic philosophical constraints, it's probably the best they can do. I find it a bit ironic that today's science doesn't know with certainty exactly how the universe came to be, but it knows with certainty that God wasn't involved. To repeat Chesterton's point, they reject the idea of God creating everything out of nothing but are comfortable with the idea that nothing created everything.

Not every scientist "believes" in the Big Bang theory, but most scientists committed to Darwinism do. Even among scientists who believe in the Big Bang theory there are a lot of differing opinions about various aspects of the theory. At best, the theory has a number of technical assumptions that serve as Band-Aids to keep the theory alive when conflicting observations and data threaten it.

For starters, the creation of our vast universe with at least 200 billion galaxies, each with hundreds of billions of stars by a quantum fluctuation out of nothingness isn't a terribly satisfactory explanation for the origin of the universe. Science can't say where all this came from or why it happened. Earlier I pointed out that if you torture data long enough you can make it say anything. The same thing is true in theoretical physics. If you must explain everything out of nothing by unknown *natural* processes without a creator, it seems the theories are more incredible than a creator.

Soon after the Big Bang theory was proposed it had to be modified with a cold whoosh before the hot big bang in order to explain certain data. This is called the inflation theory and it is one of the bandages tacked onto the Big Bang theory to keep it alive. Some scientific models claim the cold whoosh of inflation could not stop all at once, so it had to stop here and there and continue elsewhere leaving voids in space which became new universes. Of course, we are only aware of one universe, but we are darn lucky to be in the one where life "evolved."

More recently science has added unobserved dark matter and dark energy to keep the Big Bang alive. Dark matter is inferred in part by unexplained gravitational forces. It is not observable but is estimated to greatly exceed the visible matter in the universe. There are many things the Big Bang theory can't explain, and it is held together with dark matter, theoretical physics, imagination and assumptions (bandages). The Big Bang model assumes that matter was distributed evenly throughout the universe and the cosmic microwave background (CMB) radiation should be at an even temperature throughout the universe. The CMB is cited as evidence of the Big Bang but it turns out that matter is not evenly distributed throughout the universe and the CMB radiation temperature is not the same everywhere.

The Big Bang is a mathematical model attempting to explain the current universe based on unobserved and unrepeatable past events, naturalistic assumptions, theory and present-day observations. The Big Bang model isn't able to explain many observations, but it survives because it is natural and avoids a creator/designer. Some galaxies in what is considered the "young" part of the universe are too large to be compatible with the Big Bang model. Physics demands that a hot Big Bang should create equal amounts of matter and anti-matter but only traces of anti-matter are found. The Big Bang model should have created a large number of population III stars, but none have been found. The problems with the Big Bang model are numerous so it's no wonder some scientist don't accept the Big Bang theory even though they are committed to a naturalistic origin of the universe. I suspect the Big Bang theory is something like a place holder until another naturalistic model arises to explain everything out of nothing.

I am of the opinion that an omnipotent creator God is a superior explanation for the origin of the universe. Our universe has been created and designed with laws, properties and constraints that fall within extremely narrow ranges to support the existence of atomic particles as we know them and life. It is highly unreasonable to attribute the design and order of our universe to a Big Bang and a lot of luck. We are clearly the product of a big God not a Big Bang.

For the person wanting to know and understand weaknesses of the Big Bang theory at a more technical level, I have provided some suggested resources below. You are not likely to find weaknesses in the Big Bang and Darwinism in public school textbooks or popular newspapers and magazines as they are generally committed to defending the Darwinian faith

FOR FURTHER STUDY

CREATION WEBSITES

www.icr.org
www.answersingenesis.org
www.trueorigin.org

INTELLIGENT DESIGN WEBSITES

www.arn.org
www.discovery.org/id/

Big Bang article on Answers In Genesis website: https://
answersingenesis.org/big-bang/

CHAPTER EIGHT
HOW THE LAWS OF THERMODYNAMICS RELATE TO ORIGINS

Don't panic! I'm guessing most of you haven't used the word "thermodynamics" in a conversation in some time. Remember, this is the layman's edition, and it is *easy* to understand.

Now that we have the universe in place, it's probably time to think about the laws of thermodynamics. If you haven't been thinking about the laws of thermodynamics recently, no worries. It can get highly technical, but the basic concepts are easy to understand and so very important to the creation-evolution debate.

THE FIRST LAW OF THERMODYNAMICS is known as the law of conservation of energy. The law holds that energy cannot be created nor destroyed. Matter is a form of energy (according to Einstein's equation of equivalence $E=mc^2$) so the total mass and energy of the universe is fixed. Matter can be converted to energy and energy to matter but the total mass and energy in the universe is fixed and constant.

THE SECOND LAW OF THERMODYNAMICS holds that the entropy in a closed system (such as the universe) tends to increase over time. This description of the second law is a little confusing to someone who hasn't studied the matter but it's not a difficult concept and readily observable in our everyday experience. Entropy in thermodynamics talk is a measure of the unavailability of thermal energy to do mechanical work. You can think of it this way. If you have a kettle of boiling water on the stove and you turn off the burner under the kettle the burner cools (entropy increases) and the available heat can no longer boil the water to cook your eggs. In a similar way, the total available energy in the universe is decreasing over time. You could say the universe is cooling down after a supposed hot big bang. At some point in the distant future, the universe will reach a point (sometimes called heat death) where there is no available energy to make things happen. The total mass/energy will not change but it will be in equilibrium and there will be no available energy to drive processes. Life would not be possible in this state. (Don't worry, it's going to take a while. If you insist on worrying, skip to chapter nineteen.)

Now, let's look at the laws of thermodynamics as they apply to the origins debate. The first law says that mass and energy cannot be created nor destroyed, and the total mass and energy of the universe is fixed. This is a bit awkward for the Big Bang model where it is claimed the entire mass/energy of the universe popped into existence in defiance of the first law. If Darwinists maintain that it was a product of some unknown natural law, it seems reasonable to ask how there could be natural laws at work in the nothingness that existed before there was a universe? Just asking.

The second law of thermodynamics tells us that the total available energy in the universe is running down. The universe began with a point of maximum available energy and that available energy has been diminishing but, lucky for us, has not yet reached equilibrium where there is no energy available to make things work. All suns will burn out and all radiation will stop. The universe will still exist, but nothing will be happening. It will be a cold still universe. Again, we see that the universe had a beginning in time….it is not eternal. What energy source or first cause, apart from the universe, is capable of creating the universe in time and space?

The principle of entropy in the second law of thermodynamics extends beyond loss of available energy to apply to the state of order in a closed system. Not only does available energy decrease in a closed system, so does order. We see this law at work in our everyday experiences. Over time, things run down, wear out, break and die…. order decreases. When you buy a new car, or a house does it stay new forever? It doesn't. Over time it runs down, wears out, breaks and dies. It takes a lot of maintenance and upkeep to keep a car or house or anything in functional order. If you look into a mirror over time you will see entropy at work. Living things run down, wear out, break and die.

I think the innate tendency for many people to believe in creation relates to our real-world observations of life and nature, the laws of thermodynamics at work. Things don't tend to self-organize or get more organized and complex over time. In the real-world things go the other direction. This is increasingly obvious when we discuss biology and the origin of life.

Life exceeds chemistry in the same way that information in a book exceeds ink and paper. From the perspective of natural chemistry, life is highly *unnatural.* The complex molecular components of biological systems and the systems themselves do not arise in nature even if all the right chemical elements are present and the boundary conditions optimum. Life is dependent on *information* and *organization* that is *imposed* on rather than *inherent* in the chemistry on which it relies. As a violin doesn't compose the music it plays, the chemical elements within living organisms dance to a tune composed and arranged outside of and apart from the chemistry itself. When the music stops, and chemicals react in ordinary ways we see death and decay….the second law prevails.

Watch a dead cat on the side of a road for a few hot summer days and you'll see natural unguided chemistry at work. All of the chemicals required for life are present and the sun is providing an energy source but when chemicals act *naturally,* they don't *evolve* into increasingly ordered complex systems, they return to dust. In accordance with the second law of thermodynamics, when the support systems (respiratory, circulatory, etc.) cease to function, the chemicals of which the dead cat consists, react in their *normal* unorganized way. These normal chemical reactions tend toward increased entropy or a state of maximum stability in equilibrium. Information, order and complexity are overtaken by simplicity, disorder and randomness. From a scientific perspective we can say that dead cats tend to remain dead and to decay. From a subjective human perspective, we can add that they tend to stink a lot in the process. From an evolutionary perspective we can say that dead cat evolution is going in the wrong direction.

When we consider the first and second laws of thermodynamics, whether from a scientific perspective or from our daily experiences and observations, we don't see them supporting an entirely naturalistic origins model. I find an omnipotent creator God is a better explanation for the origin and order of the universe than an explosion out of nothing. Is that an unscientific statement or a non-scientific statement? If it's what happened, how could it not be science? The word science is derived from the Latin word *scientia* which means "knowledge." When our public education system and popular media conflate and define the discipline of science with philosophical naturalism, they are limiting their knowledge to what they want to believe. In the name of science, they are trashing scientific objectivity. Worse yet, they are trying to impose their belief system on others.

FOR FURTHER STUDY

CREATION WEBSITES

www.icr.org
www.answersingenesis.org
www.trueorigin.org

INTELLIGENT DESIGN WEBSITES

www.arn.org
www.discovery.org/id/

CHAPTER NINE
IS THE ORIGIN OF LIFE A MATTER OF CHEMICALS AND CONDITIONS?

The origin of life is exceedingly important in the creation-evolution debate. The Darwinist position holds that all the diversity and complexity of life on earth has evolved from an event where unguided chemical elements came together to form a "simple" living creature, likely a single celled organism. The naturalistic evolutionary belief system insists that, given the right chemicals and conditions, life will spontaneously spring forth. Remember my college biology textbook stated this unequivocally as "*...once the earth had originated in its ancient form, with particular chemical and physical properties, it was then virtually inevitable that life would later originate on it also.*" For the Darwinist, life is a simple matter of chemicals and conditions.

Obviously, living organisms can't evolve into the diversity we see in the fossil record and living today until you first have life. The origin of the first living organism is a necessary first step. It is a very problematic first step for Darwinism.

The theoretical original "simple" single celled organism must have the ability to ingest and process nutrients and replicate itself, otherwise the first living organism would have been the last living organism. The reality is, there is no such thing as a "simple" living cell. This is a lot to ask of unguided chemicals, but it must happen before biological evolution can begin. The evolutionary mechanisms of natural selection, mutation, and survival of the fittest have no role before there is life. No change is advantageous to a rock or non-living chemicals. All the systems to sustain and replicate life must come together at the same time and place and be constrained within a protective membrane in order for life to begin.

A couple hundred years ago, it was commonly thought that life could spontaneously spring forth out of non-living elements. It was known as SPONTANEOUS GENERATION. It was readily observable because flies could be seen to arise out of swamp muck, maggots could be seen to spring forth from rotting meat and a bag of grain would often produce mice.

Louis Pasteur conducted experiments that disproved spontaneous generation and proposed the LAW OF BIOGENESIS which holds that life only comes from pre-existing life, not from non-living material. To this day, life has never been observed to come from non-living material and life has not been created in a laboratory. Even if scientists were to create life in a laboratory it would demonstrate intelligent design, not spontaneous generation.

Well, nuts! The modern theory of evolution is based on naturalistic or materialistic philosophy that *requires* life to arise from non-life by some unaided natural process. In the real world it is not happening but from the Darwinist perspective, it *must* have happened, and the proof they have is that we are here, and God didn't do it. Well, that's obviously not proof; like my college biology textbook, it's a statement of faith.

Today, the term spontaneous generation has been replaced with ABIOGENESIS. (A – without, BIO – life, GENESIS = origin) Spontaneous generation had been disproved but the concept was *resurrected* with a new name, abiogenesis. See, evolutionary "science" isn't hard once you know how it works.

In order for human life to exist, non-living chemicals must come together to form life and then evolve through increasingly complex organisms until the first human comes into existence. I've heard this journey from non-living chemicals to people described as, "goo to you by way of the zoo." The big problem is that, somehow, a chemical goo in the distant past must produce life and it's just not that easy.

I remember, at one-point years ago, being told that the origin of life might have been kick-started when a lightning bolt (an energy source) struck a warm pond filled with all the right chemical elements to form life. The general public is led to believe that life is just a matter of chemicals and conditions and the pond and lightning represent a simplistic statement of that belief. It turns out that this "once upon a pond" story could not have led to the origin of life. The building blocks of life are extremely unstable and the conditions for their forming and surviving to form significant biological molecules are highly improbable.

We have to get rid of the pond because the presence of liquid water inhibits certain chemical bonds that are necessary for life. The presence of an energy source, be it lightning or sunlight can facilitate certain chemical bonds but it also breaks down many chemical precursor molecules required for life. The presence of oxygen is also problematic for the origin of life as it also breaks down biological precursor molecules.

A chemical soup with all the right elements and none of the wrong elements for the origin of life remains the hope of Darwinism. The proposed catalyzing energy source necessary to jump-start chemical reactions to form the molecules of life have included lightning, sunlight, volcanoes, meteors, and deep-sea thermal vents. To date, scientists have performed many experiments using various chemicals and conditions to see if they could create life. No luck so far so it might not be as "inevitable" as my college biology text claimed.

Back in 1952 a Scientist by the name of Stanley Miller set up an experiment to see if natural processes could produce the basic building-block molecules necessary for life. To do this, he filled a glass chamber with various gases that were assumed to be capable of creating necessary biological molecules when he shot sparks into the chamber. The sparks represented lightning or an energy source. The gases in the chamber included methane, ammonia, hydrogen, and water vapor. There was no oxygen in the chamber because oxygen tends to rapidly oxidize and destroy DNA, RNA, proteins, and cell membranes that are necessary for life. It was assumed, without evidence but for the sake of the experiment, that the early earth atmosphere contained these elements and conditions. Geologic findings indicate that oxygen was present in the early earth atmosphere and many scientists today dispute the early earth had a low or oxygen free atmosphere. Not only that, scientists today dispute the idea that the gases in Miller's spark chamber represented the early earth atmosphere in any way.

Miller's experiment to show that important biological molecules could form under natural conditions was actually conducted under most unnatural designed and manipulated conditions. The products of the experiment had to be rapidly trapped out of the spark chamber because the conditions there would quickly destroy them. The experiment created carbon monoxide and nitrogen gasses and a lot of goo which was somewhat toxic, but it also created a few amino acids.

Amino acids are the building blocks of proteins and most organisms are largely made up of proteins. Miller's experiment was proclaimed in headlines as having demonstrated how the building blocks of life and ultimately life, itself could form under "natural" conditions. The Miller experiment is still cited today, and I have encountered people who think Miller's experiment all but created life itself.

In reality, Miller's experiment was a product of intelligent design rather than natural processes and in no way represented early earth conditions. The experiment did produce some amino acids which are sub-components of proteins, but how significant is that? In the next chapter we will look at the relationship between amino acids and proteins.

FOR FURTHER STUDY

CREATION WEBSITES

www.icr.org
www.answersingenesis.org
www.trueorigin.org

INTELLIGENT DESIGN WEBSITES

www.arn.org
www.discovery.org/id/

CHAPTER TEN

FROM AMINO ACIDS TO PROTEINS

The relationship between amino acids and proteins is somewhat like the relationship of a word to a paragraph. A word by itself cannot carry the message of a paragraph and an amino acid molecule by itself cannot provide the function of a protein. The words in a paragraph must be organized in a highly specific way for the message of the paragraph to be conveyed. In a similar way, specific amino acids must be linked in a specific way to create a functional protein.

While Darwinists like to portray the origin of life as beginning with a simple cell, the fact is there is no such thing as a simple living cell. Life is complex and dependent on order and specificity. Random processes cannot create the order and specificity required for life. In the Intelligent Design camp, William Dembski came up with the term SPECIFIED COMPLEXITY. This is a useful concept to understand in biological systems. A thousand random letters on a page is complex but it carries no information. When those same letters are arranged into words in a specified order there is still complexity but there is also meaning and information. To be sure, biological systems are highly complex, but they are also highly specified. An explosion in a print shop results in extreme complexity but without specificity. You won't find a sonnet arranged out of the randomness of such an explosion. In the same way, a random collection of 150 amino acids will not form a functional protein. A protein gets its functionality out of the specific amino acids and the order in which they are arranged. Specificity and order are the key to biological life.

Amino acids are a particular type of biological molecule. Like most biological molecules they consist of six elements (carbon, hydrogen, oxygen, phosphorus, sulfur and nitrogen). Many molecules can be formed from these basic elements and an amino acid is just one such molecule. Amino acids are not alive, but many of them are building blocks of proteins which are important biological molecules. Proteins are complex molecules, but they aren't alive either. I like to say a protein is to a living organism as a brick is to a hospital. A brick is an important part of a hospital, but it doesn't design rooms, windows, doors, electrical and plumbing systems, etc.. We are largely made of proteins, our skin, hair, organs, muscles, etc. are made from different kinds of proteins; they are building blocks and enzymes that make life possible.

So, we are in large part, made out of proteins and proteins are made of amino acids. Amino acids have two parts, a backbone and a side chain. All amino acids have the same chemical backbone structure, but different kinds of amino acids have different side chains. The side chain is a unique combination of atoms that give each different amino acid its unique electro-chemical properties. This side chain extends off the left or right side of the common backbone structure. This means that while there may be hundreds of different kinds of amino acids, each kind of amino acid has two versions, a left-hand version and a right-hand version. Each version has identical properties except for the position of the side chain. Visualize this by holding your open hands in front of you and see that they are identical except reversed and the thumb comes off a different side.

When amino acids form, they form in equal quantities of their left and right-handed varieties. This is called a racemic mix. Right now, you are probably wondering why I'm telling you more than you ever wanted to know about amino acids. Hang with me and I will show you why this is significant. While there are easily over 100 different amino acids that can form, only 20 specific amino acids are used to make proteins in living organisms. Not only that, but only left-handed amino acids are used to make proteins.

If you are tracking with me, you now know that six chemical elements can form over 100 different kinds of amino acids. You know that these amino acids always form in equal quantities of left and right-handed models. You also know that only the left-handed version of 20 specific amino acids are used to form proteins in living organism. You are just beginning to see the critical role specificity has in the building blocks of life.

In order to form a functional protein, amino acids are linked together in a chain. The chemical bond between amino acids in the chain are called peptide bonds making what is called a polypeptide chain. This chain can consist of 50 to more than 1000 amino acids. Each amino acid in the chain must be a left-handed version and one of the 20 amino acids that form functional proteins. But wait, this isn't complicated enough yet. It turns out that in many, if not most, cases each link in the chain of up to 1000 amino acids must be a specific one of the 20 amino acids. This is important because a chain of amino acids is not yet a functional protein. The chain must fold into a specific three-dimensional shape and it's that three-dimensional shape that gives each protein it's function. The unique electro-chemical properties and precise position of each amino acid in the chain controls how the chain of amino acids will fold into a functional protein. Sometimes, more than one amino acid will work in a particular position but sometimes one amino acid in the wrong sequence will prevent proper folding and thus destroy the functionality of the protein.

We know that each amino acid in a linked chain of amino acids must often be a specific amino acid with the right electro-chemical properties required for proper folding and ultimate functionality of the protein it is forming. If we compare amino acids in a protein to words in a sentence, I can give a rough example of how specificity is necessary.

Consider the sentence: "I WILL GO FISHING." If words are randomly inserted or removed (like amino acids randomly inserted or removed in a protein) the meaning of the sentence (function of the protein) can be lost. If the word WILL is replaced with WON'T the meaning is changed. If the word NOT is inserted after WILL, the meaning is changed. If FISHING is replaced with HUNTING, the meaning is changed. I could go on but I'm sure you have the idea. Specificity in word placement in a sentence is necessary to carry the intended meaning. In the same way specificity is required in amino acid placement in order to form a functional protein. Keep in mind my example sentence was only four words long but there can be over a thousand amino acids in a protein. Without specificity there is loss of meaning and loss of function. Random, unintended, uncontrolled, unguided processes can produce complexity but not specificity, meaning, order and function.

For some complicated proteins, helper molecules (called chaperone proteins) must assist the folding process. Isn't it handy that random chance provided helper molecules to assist a chain of amino acids to fold into a specific shape to make a functional protein? If it takes proteins to make functional proteins, which came first? How does a chaperone protein know which proteins need help to fold properly and what the final shape must be for functionality? Is it reasonable to believe this order and functionality is the natural but unintended product of an expanding cloud of mostly hydrogen gas after a distant Big Bang?

If specificity and order are essential for functional biological molecules, how does this relate to the second law of thermodynamics? In the natural world, in accordance with the second law and our everyday experience and observations, randomness prevails. Things run down, wear out, break and die. They don't trend toward specificity and functional order. What role does chance play in biological systems?

A wide variety of highly specific proteins are required for life. In the next chapter we will look at the probability of functional proteins coming into existence by random chance.

FOR FURTHER STUDY

CREATION WEBSITES

www.icr.org
www.answersingenesis.org
www.trueorigin.org

INTELLIGENT DESIGN WEBSITES

www.arn.org
www.discovery.org/id/

CHAPTER ELEVEN
PROTEINS BY CHANCE – PROBABLY NOT

By now I hope I have demonstrated that when it comes to living organisms in general and proteins in particular, we are not just dealing with complexity. We are dealing with specified complexity. Let's consider what we know about the formation of functional proteins to see how impressive the Miller spark chamber was when it produced only a few of the 20 amino acids required to build functional proteins. Bear in mind that Miller's experiment didn't produce all of the 20 amino acids required for functional proteins but worse, the amino acids it produced were in equal quantities of left and right-handed versions. The insertion of one right-handed amino acid would destroy the functionality of any prospective protein. Miller's experiment used intelligent design in an attempt to demonstrate how natural unguided processes might produce molecules leading to a living organism. (Please re-read the last sentence if you missed the irony.) While the Miller experiment produced some of the amino acids found in proteins, it produced even more of the wrong kinds of amino acids that would destroy protein functionality.

What would it take for blind chance to form a functional protein? Have you ever flipped a coin to resolve an issue? Since a coin has two sides, a head and a tail, each time you toss a coin either side has a 50% chance of landing up. Put another way, we could say that since a coin has two sides with only one side as the "head," the probability of a coin

"

landing with the head up is ½. A die (singular for dice) has six sides so the probability of any particular number landing up is one in six or 1/6.

Now, if you wanted to jointly flip a coin to get a heads and throw a die to get a six, your odds of achieving this in one try would be determined by multiplying the coin odds (½) times the die odds (1/6). Your odds would then be 1/12 or one chance in twelve attempts. You wouldn't necessarily get the heads and the six in 12 tries, but the odds are you could get it once in 12 tries.

Where are we going with this probability stuff? Let's look at another example and perhaps you will see where we're going. Michael Stubbs (B.Sc., B.G.C.E., NM. Ed) in an article published in Creation Magazine titled "Evolution, Chance and Creation" http://www. answersingenesis.org/home/area/magazines/docs/v42dice.asp, provides an excellent illustration of probability involving specified complexity. In his example, Stubbs calculates the probability of spelling the phrase, "THE THEORY OF EVOLUTION" by random chance selection and sequencing of letters and spaces in the correct order.

The phrase, "THE THEORY OF EVOLUTION" contains 23 letter/space slots or components. There are 26 letters in the alphabet and when we add one space, we end up with a total of 27 elements to select from each time a letter or space is to be sequenced into the phrase. Since the rules of spelling, syntax and grammar specify that only one correct letter/space element is correct for each of the 23 slots in the phrase, we would say that we are dealing with *specified complexity*. Each time one of the 27 letter or space elements is to be randomly sequenced into one of the 23 slots in the phrase, you have one chance in 27 of getting the correct letter/space element. This computes to one chance in $(27)^{23}$ which is another way of saying one chance in *six hundred million trillion attempts*. Stubbs puts this large number into perspective by pointing out that if you could remove and replace letters at the rate of one billion times per second, you could spell out the phrase "THE THEORY OF EVOLUTION" only once in 26,000,000,000,000 (26 quadrillion) years. Since the earth is supposedly about 5 billion years old (by standard evolutionary time reckoning) we will have to wait 5 million more earth-ages for chance to spell out "THE THEORY OF EVOLUTION."

An important point to make here is that the specified complexity in the phrase "THE THEORY OF EVOLUTION" is infinitely simpler than the specified complexity in the simplest life form. As Don Batten, PhD points out in his an article titled "Cheating With Chance" published by www.answersingenesis.org, *"The probability of the chance formation of a hypothetical functional 'simple cell' given all the ingredients, is acknowledged to be worse than 1 in 10^{57800}. This is a chance of 1 in a number with 57,800 zeros. It would take 11 full pages of magazine type to print this number. To try to put this in perspective, there are about 10^{80} (a number with 80 zeros) electrons in the universe. Even if every electron in our universe were another universe the same size as ours, that would only amount to 10^{160} electrons."*

Now let's look at the probability of a single rather simple protein evolving by chance in a natural environment. To avoid making this too complicated we will make a number of assumptions that are not remotely realistic but they help us keep things simple so we can focus on the probability issue.

– ENVIRONMENTAL ASSUMPTIONS: We will assume the existence of an ideal chemical and temperature environment for our protein to evolve in. Certain chemical and temperature boundary conditions are required in order for biological molecules to form and survive for any period of time. We will assume, for purposes of the argument, that all of these conditions are ideal.

– The presence of free oxygen or liquid water would precipitate many wrong reactions that would preclude the formation and/or longevity of important biological molecules. Amino acids degrade in the presence of oxygen and peptide bonds (the bonds between amino acids) come apart in the presence of water. For this reason, evolutionists like to assume that life evolved within a special chemical environment, a primordial soup, and a special "reducing atmosphere with no free oxygen.

– I should point out that the absence of oxygen implies the absence of ozone (a form of oxygen) that protects the earth from UV light. This is a problem for evolution because UV light destroys biological molecules. Let's assume that some other mechanism is available to filter out UV light.

– Even though there is no positive evidence in the geologic column that these special conditions required for chemical evolution ever existed on earth and much evidence that denies the existence of these conditions, we will grant that by some miraculous accident of nature, all of the environmental conditions necessary for chemical evolution exist - for the sake of the argument.

– AMINO ACID ASSUMPTIONS: For the sake of the argument we will also assume that somehow the conditions were right to cover the earth with an "ocean" of amino acids. We will assume that somehow something happened that removed all of the *wrong* amino acids. Remember, there are over 100 different kinds of amino acids but only 20 of them are used to build proteins. If you were to add one of these wrong amino acids to a growing peptide chain you would destroy the functionality of the protein. To remove this possibility, we are going to assume that somehow all of the wrong amino acids have been removed so they can't foul up the works. We will assume our ocean of amino acids consists of *only* the 20 *correct* amino acids to form functional proteins.

Now with these assumptions in place, let's see how a functional protein might form by chance under these (unrealistic) ideal conditions. A completed functional protein would consist of somewhere between 50 and 1000 or more amino acids. To make our protein, we must first select only left-handed amino acids and then select the correct one of 20 possible amino acids to fit in the proper sequence in the peptide chain that will become our protein. Remember, the sequence of amino acids is highly specific (specified complexity) in order to ensure the protein is able to fold into the 3-dimensional shape required for functionality. Remember also, that amino acids come in a 50/50 (racemic) mix of left and right-handed forms so each time one is selected to add to the growing chain of amino acids, you have a 50% chance or ½ probability of obtaining the required left handed form of the molecule. If we assume a relatively small protein consisting of 400 amino acids, the chance of obtaining 400 left-handed amino acids in a row would be like flipping a coin 400 times and getting a "heads" each time. But that's the easy part.

Each time you select a left-handed amino acid, it must be one of the 20 right kinds of amino acids. We have assumed an ocean of only the correct kind of amino acids but we must still deal with the fact that in many, if not most cases, only one of the 20 is the correct amino acid to provide that particular location in the amino acid chain with the proper electro-chemical properties to allow the chain to fold into the proper 3-dimensional shape required for functionality. In other words, you have 1 chance in 20 or 1/20 chance of getting the correct amino acid each time you select one at random to fill a link in the protein peptide chain. All of this means that each time you add an amino acid to the growing protein chain you have ½ chance of getting a left-handed one and 1/20 chance of getting the right amino acid. This computes to ½ X 1/20 or one chance in 40 (1/40) of getting the correct amino acid for the slot in the peptide chain. If a particular protein consists of 400 amino acids, you would determine the probability of it "evolving" by chance by multiplying 1/40 X 1/40 X 1/40 X 1/40, etc…400 times. The first time you get a right handed amino acid or the wrong one of the 20 amino acids in the wrong slot in the sequence, you have to start over at the beginning because the protein is now "mutated" and would not be functional.

It has been discovered that sometimes, more than one of the 20 amino acids has the right chemical composition to work in a particular position in an amino acid chain without destroying the functionality of the protein. However, sometimes the wrong amino acid in a single position will destroy the functionality, so specificity remains critical.

Joseph Mastropaolo PhD in an article titled "Evolution is Biologically Impossible" published in www.icr.org, quotes Hubert Yockey, an evolutionist, as computing the probability of the natural evolution of one small common protein as one chance in 2.3 times ten billion vigintillion. Ten billion vigintillion is a one followed by 75 zeroes. Mastropaolo goes on to say, *"..to put it in evolutionary terms, if a random mutation is provided every second from the alleged birth of the universe, then to date that protein molecule would be only 43% of the way to completion."*

I should point out that we consist of 100,000 or so different kinds of proteins. Even if one needed protein could evolve by chance, it doesn't *know* how to form a living creature. Remember, a functional protein is like a brick in a building. The design for the building doesn't reside in the brick.

As all of this pertains to the origin of life, I hope you can see that the fact that Miller's spark chamber experiment produced a few amino acids is insignificant in the grand scheme of things. Miller's spark chamber experiment was based on some wrong assumptions, but it was a legitimate attempt to discover a naturalistic path to the origin of life. In reality Miller's experiment had very little to do with the origin of life. It demonstrated how amino acids could form under the right conditions, but it didn't demonstrate how amino acids could form complex highly specific functional proteins.

To this day, scientists have been unable to demonstrate how life could have originated through natural processes. The specificity required for many biological molecules such as proteins and DNA make it statistically impossible for life to happen by chance even if you grant all the time available since the "creation" of the universe.

There is much more that could be said, and examples given but I have demonstrated that natural laws (thermodynamics and chemistry) and statistical probability preclude the naturalistic origin of complex proteins. Even if highly specific functional proteins could evolve by chance, they are not alive, they are just building blocks of organs and biological systems, bricks in a hospital. Protein building blocks don't know how or where to build a heart or a brain. Living organisms consist of many different complex proteins so the origin of life by natural processes and chance is impossible many times over.

Today in the newspapers and science journals we don't hear as much about the origin of life as we do of the evolution of life. A paleontologist finds a bone somewhere and the articles say, evolution goes this way. Another paleontologist finds another bone and it changes everything and evolution is said to go another way. Darwinists have not lost faith in abiogenesis, but the origin of life remains a problem.

The intuitive belief many of us have in creation/design as opposed to the naturalistic origin of life is supported by natural laws and probabilities as well as our everyday observations. In accordance with the second law of thermodynamics and our daily observations, we see that, over time, things don't trend toward self-organization or specified complexity, they break down and fall apart – like the dead cat under the hot sun. In accordance with the second law of thermodynamics and our experiences in life, things do not tend to integrate, they disintegrate. The amazing complexity of living organisms is the result of intricate design and organization imposed on chemistry.

The chemical elements of which living organisms consist have certain electro-chemical properties, but they don't contain the information necessary to self-organize into complex functional proteins or a living creature. The integrated structures and systems in living organisms, like various organs, the cardio-vascular system, respiratory system, immune system, blood clotting system, etc. are the product of information and design. Information and design are not properties inherent in carbon, hydrogen, oxygen, phosphorus, sulfur and nitrogen. While we consist largely of these basic elements, these elements don't have the capacity to anticipate and design.

These six elements (and a few others) are arranged to store and use massive amounts of information in living organisms. However, chemistry and matter cannot originate the information they can contain. The information that enables life is not inherent in matter, it is imposed on matter. You are alive because your chemistry dances to the tune of vast amounts of information. Life depends on information so where does the information come from? What do we know about information – let's examine that in the next chapter.

FOR FURTHER STUDY

CREATION WEBSITES

www.icr.org
www.answersingenesis.org
www.trueorigin.org

INTELLIGENT DESIGN WEBSITES

www.arn.org
www.discovery.org/id/

Evolution, Chance and Creation article in Creation Magazine http://www.answersingenesis.org/home/area/magazines/docs/v42dice.asp
Cheating With Chance article by Don Batten, PhD published by www.answersingenesis.org,
Evolution is Biologically Impossible, article by Joseph Mastropaolo PhD published in www.icr.org,

CHAPTER TWELVE
LIFE REQUIRES INFORMATION

If you recall, my college biology textbook made the claim that life was an "inevitable" result of chemicals and conditions – and, of course, natural processes. It turns out that "inevitable" doesn't stand in the face of real-world chemistry and statistical probabilities. After much study, I know today that while this statement was in a science textbook, it was a statement of faith not of science. To this day, after much research and many experiments, science has been unable to create life by manipulating chemicals and conditions. Life, at any level, is extremely complex. Something more than chemicals and conditions is required. That something more is information and information is not inherent in matter or chemistry.

What is information? Information is not a physical thing. You can't weigh it or see it, but you use it every day. Information is not inherent in matter, but matter can transmit and store information. Let me give you an example. If you are reading this in a book, you are receiving and (hopefully) processing information. The information isn't in the paper or the ink is it. The information has been imposed on ink and paper by an intelligent source. (We can argue about just how intelligent the source is on another day.) The same thing applies if you are reading this on a computer or electronic device. The information isn't in electricity or light or bits and bytes, it is imposed on these material elements by a mind. The software in your computer is like information, it has no mass. You can weigh your computer with all your software programs, documents, photographs and music then remove all the data encoded in your software and the weight won't change. The information content in software has no mass and is not physical but it can be encoded in physical matter. Information isn't a product of random events; it originates from an intelligent source.

Information theory is a branch of mathematics. Initially, the science of information was just looking at probabilities such as how a random process might arrange 26 letters of the alphabet into a word. For example, what is the chance of an explosion in a print shop spelling antidisestablishmentarianism? The probabilities of the chance formation of words is one thing but over time information theory began to look beyond probabilities to messages and meaning. Words and sentences and books have meanings and purposes that transcend the probabilities of letter arrangement.

I have read a number of articles on information and information theory, but I recommend a book written by a German information scientist by the name of Dr. Werner Gitt. The book has been translated into English and is titled "In The Beginning was Information." I found the book very interesting but a little painful to read. Dr. Gitt writes as a scientist, meticulously unfolding the elements of information theory. It's well worth the struggle to get through the book as you will have a good handle on information theory when done.

Information theory can be complex, but the basic concepts are easily understood. Let's look at some of the properties of information. We have mentioned some but there is more:

1. **Information is immaterial.** it has no physical properties; it is a mental construct.

2. **Information originates from an intelligent source, a mind.** It is not inherent in matter and not the result of random processes or events.

3. **Information cannot originate in matter but can be imposed on or transmitted by material elements.** It can be ink on paper, bits and bytes in computer code, smoke signals, light, radio waves, sound waves, etc..

4. **Information is represented or transmitted by a code that is understood by the sender and a receiver**. The code can be letters of an alphabet, it can be a two unit code like the dots and dashes of the Morse code, it can be the four unit code of DNA, it can involve all sorts of symbols but it includes some sort of syntax (order) and semantics (meaning).

5. **Information represents or substitutes for something else**. This last concept is a little difficult to explain but I can give an example. If I use the code symbols of the letters in the English language to spell CAT, you know I am talking about a particular animal. These three letters of the 26-letter alphabet, in this order spell and represent a particular animal. The letter-word CAT isn't the real thing, it merely represents the real thing, so you don't have to put it outside before going to bed. The meaning is found in the specific order of these three letters of the 26-letter alphabet, so specificity is critical to meaning (semantics). If I were to change the order of these same three letters to spell ACT or TAC, the meaning would be entirely different. However, because I the sender, and you the receiver, are using the same alphabet code (item 4 above), information is transmitted.

If I were writing in Spanish or Chinese or any other language, the code (letters or symbols) would change but the message or information would still be represented and understood by the targeted receiver.

I used the three-letter word CAT to keep things simple for the illustration. I suppose we could imagine how an explosion in a print shop might possibly produce the word CAT in the mess of random letters lying about on the floor. But what's the probability of the explosion creating the sentence and punctuation in in this sentence... *How did the CAT ACT when it stepped on the TAC?* The letters, spaces, and punctuation in this specific sequence (syntax) form words, each with independent (semantic) meaning but when arranged in this order they carry much more meaning or information than they do independently or individually.

Today, Information science shows us that complex information doesn't come out of random processes. We know this intuitively and the second law of thermodynamics explains it. However, I remember back in my college days there was much confidence that random processes could produce complex information. There was a common saying that claimed that an infinite number of monkeys at an infinite number of typewriters would invariably produce the complete works of Shakespeare. My biology textbook referencing the origin of life said it a little differently but was just as wrong.

Hopefully, I've given you a rudimentary understanding of information theory and you understand that matter doesn't originate information, but that information can be imposed on matter by an intelligent source. Ideally, I have just reinforced what you already know from experience and intuition. In the next chapter, let's take a look at how information relates to biological systems and the origin of life by natural processes.

FOR FURTHER STUDY

CREATION WEBSITES

www.icr.org
www.answersingenesis.org
www.trueorigin.org

INTELLIGENT DESIGN WEBSITES

www.arn.org
www.discovery.org/id/

In The Beginning Was Information, by Werner Gitt, published by Master Books

CHAPTER THIRTEEN
INFORMATION'S ROLE IN BIOLOGY

Previously we talked about how proteins are made out of highly specific arrangements of 20 amino acids. This is an example of information in biology. The 20 amino acids can be compared to our 26-letter alphabet. Their highly specific arrangement in order to make a functional protein can be compared to the specific arrangement of the letters of the alphabet to form a proper word or sentence. A typical protein can consist of hundreds of amino acid letters but spelling (the sequential arrangement of amino acid/letters) is critical to functionality. In some cases, a single amino acid out of sequence can cause the protein to be unable to fold into the required shape for functionality. Genetic spelling is critical for life to exist. Random changes to amino acid sequences are called mutations and they can be fatal.

The recipe (instructions) for every protein is contained in coded form in our DNA. Human DNA contains about three billion code elements (letters). This is enough "letters" to write a library of over 800 large books the size of a Bible. If you have read the entire Bible you know this is significant. The information in DNA not only codes for proteins but for every organ and feature of our body. Our knowledge of DNA is continually growing and yet, it seems there is so much more to know. The complexity of design and function within DNA points to an awesome Creator.

Most people know that DNA is a shortened term for **d**eoxyribo**n**ucleic **a**cid. DNA is much easier to say and a lot easier to spell than deoxyribonucleic acid so we will just use the term DNA. DNA is a biological molecule found in almost every cell of every living organism. The DNA molecule is designed to store and carry information. Most of us know DNA is the storehouse of genetic information. (There's that word "information.")

You have likely seen pictures or drawings of the DNA molecule with its double helix shape, like a spiraling ladder. However, DNA's functional shape in a cell is much more complex than just the double helix. DNA is arranged into highly efficient structures called fractal globules which allows vast amounts of material to be packed into a small space without becoming knotted. A fractal is a mathematic set or geometric shape in which the same pattern repeats in increasingly smaller scales. DNA is the densest information storage medium known to mankind.

The DNA for every living organism may have some similarities but it is unique for every organism because it contains the information to create that particular organism. Human DNA carries instructions to assemble a human and bird DNA has the information to make a bird, etc.. You can think of DNA as something similar to computer software. It contains instructions (information) to perform functions like self-replication, or how to construct a human being with eyes, brain, heart and big toes all in the right place and functional.

The information content of DNA is found in the arrangement of a four-letter code. Remember in the last chapter we learned that information is recorded and transmitted in a code. The information in proteins is found in the arrangement of 20 amino acids, a 20 letter or component code. The four letters of DNA code are found in four chemical bases, **adenine, cytosine, guanine, and thymine**. (A chemical base is substance such as ammonia that releases hydroxide ions in an aqueous solution.) For convenience these bases in DNA are usually abbreviated as A,C,G, and T. Just as information in the two element Morse Code is found in the arrangement of dots and dashes, the information in DNA is found in the specific arrangements of the four (ATGC) code elements. ATG&C are the dots and dashes of the genetic code.

Somewhat like a protein that begins as a linear arrangement (chain) of linked amino acids that folds into a functional three-dimensional shaped protein, DNA is a line of code elements that folds into a more functional shape. It has been discovered that when the genome folds into a three-dimensional shape, it folds in a fractal pattern. Within this folded structure, the gene elements needed to work together for a particular function may not be located together in linear space but they are often located next to each other in three-dimensional space…. sometimes even on different chromosomes.

Originally it was thought that a particular sequence of DNA code, called a gene, programmed for a single specific protein. Recently we have learned that a sequence of DNA can code for multiple different proteins if the code is copied from a different starting point. We have learned that DNA actually changes shape when it is necessary to expose buried genes required to code for a needed specific protein.

In the past, it was noted that large portions of DNA don't appear to code for anything useful like proteins, so it was decided that this was junk DNA. It was assumed that junk DNA was the residue of DNA that lost its function over evolutionary time. Just as evolutionary assumptions once decided the appendix was a useless relic of evolution, so called junk DNA was assumed to be a relic of evolutionary trial and error. More recently, it has been determined that some so called "junk" DNA has structural purpose or plays other important roles in maintaining biological viability. In some cases it was discovered that mutations in "junk" DNA are harmful to the organism. It is increasingly doubtful that any DNA is junk DNA.

A marine biologist by the name of Robert Carter, PhD, gives an excellent description of some of these amazing characteristics of DNA in an interview with Del Tackett, PhD, in the DVD series titled BEYOND IS GENESIS HISTORY. I recommend the entire DVD series to anybody interested in the study of origins. (See reference below)

DNA contains switches that turn various functions on and off as needed to keep an organism functioning properly. Recently we have begun to learn more about another layer of biological code that resides on top of the genome but controls certain aspects of gene expression. This is the study of epigenetics. The term *epi* is from the Greek prefix meaning "on top of" so epigenetics is another layer of complex biological code on top of the genome (DNA). It appears that epigenetic factors are sensitive to the environment and can change gene expression based on environmental conditions such as climate, diet, stress, etc. The epigenetic language is not well understood yet, but it involves a complex set of cellular machinery that adds or removes chemical components or tags that control or modify gene expression without changing the underlying genetic code. It's a code on top of a code. Epigenetic changes can be passed on (inherited) by subsequent generations and epigenetic errors can result in disease conditions.

It should be clear that DNA, including epigenetic components, is extremely precise and complex. Random changes to DNA, called mutations, are said to be the driving force behind evolution. In real life, mutations do not add functional complexity or information to a genome. They add the type of complexity a bull adds to a china shop. Many diseases can be traced to mutational changes. If mutations were really observed to advance biology you would see Darwinists hanging out in front of X-ray machines to accumulate the "improvements" mutations would bring.

We have discussed how DNA holds vast amounts of highly specific information in coded form. The information is non-physical, it has no mass, but is imposed on and encoded into physical DNA. This DNA holds the information to specify body shape and the location, size, shape and function of every organ and biological system in every organism. It also contains the information to make every sub-component to build everything that makes life possible. Human DNA always makes a human and dog DNA always makes a dog and so on. Errors in DNA code (mutations) can result in disease, dysfunction and death, so ensuring the integrity and accuracy of the DNA code is extremely important.

DNA is very much susceptible to damage (mutational errors). Errors occur in the process of replication or with the ravages of time. Radiation and environmental factors also contribute to mutational damage. It has been estimated that we suffer thousands of mutations to our DNA every day. Without an effective DNA repair system, we would have gone extinct long ago if we could have evolved in the first place. It turns out that there are dozens of protein molecules that constantly scan our DNA and make repairs. The process is amazingly complex and sometimes different proteins act together in concert to accomplish detection and repair. These repair molecules are not perfect as some mutations survive to cause problems but without these amazing repair molecules we could not survive.

Try to imagine how the complex, highly specific DNA of living organisms evolved without a DNA repair system evolving at the same time. While you are at it, try to imagine how a molecule happens to be able to recognize and repair an error in coded information. A Darwinist would call this "apparent design." I call it amazing design, what would you call it?

I have given a very shallow and simplified description of how information relates to biological systems and in particular, DNA. The point I am trying to make is that the complex information in living organism is not and could not be the product of unguided random chance and the natural properties of matter and energy. An intelligent source, a mind, is behind information and life depends on information.

If you were to see the rocks on a hillside arranged to spell "eschew obfuscation" you wouldn't think for an instant that it was a random act of nature. You would "know" intuitively and experientially that it reflected a mind, purpose and perhaps even a tinge of humor. Yet, how utterly simple two words are when compared to the specified complexity of biological molecules, interactive codes and complex biological systems.

In the next few chapters I will underscore the evidence of design and information processing in biological systems by giving just a few examples. Many more examples could be given but I hope the information I provide will stimulate the reader to further study. As I describe biological information in the following chapters, don't get bogged down in the complexities, just ask yourself if such exquisite functional complexity could realistically be the unintended product of an explosion (a big bang). In the face of obvious design, Darwinists often use the term "apparent design." They can't deny the obvious evidence of design in living organisms, so they call it *apparent design* because they refuse to acknowledge a designer. In my opinion it takes far less faith to believe in a designer/creator than to believe life is the natural unguided and unintended product of an expanding cloud of mostly hydrogen gas after a distant big bang (cosmic flatulence?) that brought everything out of nothing.

FOR FURTHER STUDY

CREATION WEBSITES

www.icr.org
www.answersingenesis.org
www.trueorigin.org

INTELLIGENT DESIGN WEBSITES

www.arn.org
www.discovery.org/id/

Beyond Is Genesis History, DVD series www.isgenesishistory.org

CHAPTER FOURTEEN

THE MAKING OF A PROTEIN

We have talked about proteins and the precision required to produce a proper functional protein. As complex and fascinating as that is, I think the system that manufactures proteins is even more fascinating and powerful evidence of design. In the process of protein formation, a multi-component biological system must read and copy a specific section of DNA code and translate it into a functional protein. The DNA code is useless without the translating system and vice versa. How fortunate we are that the genetic code and translation system evolved simultaneously and unintentionally (by the rules of Darwinism) at the same time.

As I describe the process by which a protein is made in simplified terms, think about how this protein manufacturing system might come about as an unintended natural product of an expanding cloud of predominantly hydrogen gas after a distant Big Bang. Is there anything inherent in unguided chemistry that would cause it to form the elegantly orchestrated functional complexity you will see in protein synthesis? What does your experience, logic and intuition tell you?

The individual cells in your body are like miniature factories that are continually making things needed, like proteins, while dismantling and disposing of things not needed. In essence, every one of the billions of cells in your body is continually busy with manufacturing and maintenance to keep you alive. Of course, they do other important things like replication but let's keep this simple. Animal cells, like the cells that make up your body, are a little different than plant cells and single celled organisms but they are all complex evidence of design. I will be talking about animal (eukaryotic) cells.

Each cell is enclosed in a cell membrane which encompasses a number of smaller enclosed compartments called organelles. One of these organelles you are likely familiar with is the nucleus of the cell. All of the various organelles are suspended in a liquid called cytosol. The interior of the cell including cytosol and the organelles is called cytoplasm. All of the various organelles within the cell have important jobs to do to ensure the survival of the cell and ultimately the organism, i.e., you.

Inside the nucleus of every cell is your genome, your DNA. Your genome is something like a library with all the *information* needed to build you and every organ, system and component of every organ and system. In chapter thirteen we discussed how this information is packed into a molecule known as DNA. Your DNA directs where your head goes and where your big toe goes. It "knows" you will need a heart and lungs and how to make them, what components are needed and where they should go. It "knows" that you will need one kind of cell for your heart, another for hair, another for skin, another for brain, etc. There is a massive amount of information packed into your DNA and every cell in your body contains that information.

As discussed previously, we are, to a large extent, made of protein molecules. Our skin, hair and various organs are made largely out of different kinds of proteins. Some proteins are called enzymes and they act as catalysts to facilitate certain chemical reactions necessary for our survival. There is something like 100,000 different proteins in our body. The recipes or instructions to make all these different kinds of proteins are included in your DNA in coded form.

Let's look at how the body makes a protein it needs. Since one could probably write a book on how it's done, I'm going to do us both a favor and explain this in greatly simplified form.

The recipe for the protein we want to make is located, in coded form, somewhere in the DNA which is located in the nucleus of the cell. An

RNA (ribonucleic acid) molecule (similar in some respects to DNA) is going to enter the nucleus, locate and copy the recipe for the needed protein. The nucleus isn't going to let just any passing molecule inside, so the RNA molecule must show its chemical ID at the door. The nucleus recognizes it and allows the RNA to enter.

Once inside the nucleus, the RNA molecule attaches to the DNA molecule and moves along it until it finds the "start here" code. The RNA molecule then continues along the DNA making an RNA copy of the DNA code from that point until it comes to the DNA code that says, "stop here." Once the RNA molecule has made a copy of the recipe for the needed protein it detaches from the DNA.

Now we have an RNA molecule with a copy of the coded recipe for the protein we need to make. The sequence of DNA code for a particular protein is called a gene. We have learned that a completely different but necessary protein can be created if the RNA starts making a copy of the DNA at a different starting place. In other words, the DNA information is often overlapped in an incredible demonstration of design efficiency.

Sometimes all of the DNA code in a sequence is not needed or doesn't code for the particular protein to be built. When this happens, there are helper molecules that splice out the not-needed segment of code in the RNA copy. The not-needed code segment is called an *intron* and the remaining needed code containing the recipe for the desired protein is called an *exon*.

How fortunate we are that unguided chance mutations and natural selection produced molecules at the right time and place that "know" how and where to find and make a copy of the coded recipe for our protein. Darwinism also requires that the other molecules that "know" what to keep in the code and what to cut out of the code are products of time and chance and a heap of luck. Is it just me or does anybody else think that a code, molecules that make a copy of the code and molecules that snip out unneeded segments of code to produce a needed protein is a bit much to ask of hydrogen?

At this point we have a strand of RNA all trimmed and proper with a copy of the DNA code to make our protein. This RNA, now called messenger RNA or mRNA, is inside the cellular nucleus but it must go outside the nucleus to a special organelle (called a ribosome) that "knows" how to translate the mRNA code into our protein.

The mRNA molecule with the recipe for our protein shows it's chemical ID at the nucleus door, exits the nucleus and arrives at the ribosome translation station (organelle). The mRNA molecule then feeds through the translation station *read head* which translates the mRNA code into protein. There are steps and helper molecules involved but this is a simplified description of the process.

If you recall from chapter thirteen, the information in DNA is contained in a four-element code. The code elements are chemical bases called *nucleotides* which we abbreviate with A, T, G, and C. A set of three of these nucleotides is called a *codon* and a codon codes for a specific amino acid. Remember, proteins are made out of a chain of specific amino acids in a specific order. As our mRNA molecule with the recipe for our needed protein passes through the ribosome read head, each three nucleotides (codon) adds a specific amino acid to a growing chain of linked amino acids. The end product is a chain of specific amino acids linked in a specific order which is the initial or primary shape of our needed protein.

A large protein can consist of a thousand or more amino acids. Since it takes three nucleotides for form a codon for one amino acid, the DNA to RNA recipe for a protein can consist of three thousand or more nucleotides. All of this must be stored, copied and translated accurately to get our needed protein.

The primary shape of our protein (a long chain of amino acids) is not yet a functional protein. The chain of amino acids must fold into a specific three-dimensional shape to be functional. The folding is determined by the electro-chemical properties and sequence of amino acids in the chain. The folding is also assisted by a host of helper molecules as the protein is processed through another organelle to its functional shape.

Getting the protein processed and shipped to where it is needed within the cell is an amazing and complex process in which chemical pass codes are added or removed as the protein moves through processing stations and is eventually loaded on an enclosed transport vehicle that is equipped to dock with the organelle where the protein is needed.

The brand-new protein is deposited where it is needed, and the process is repeated in every cell every day as needed.

Think about this for a moment. We have learned that information does not originate in matter but is imposed on matter by an intelligent agent. In previous chapters we saw that information always involves a code. We saw that the "recipe" for each type of protein in our body (and much more) is stored in coded form in our DNA. What good is a coded recipe if there isn't a translation system to make use of the information? We saw that an RNA molecule makes an RNA copy of the DNA code and then uses a ribosome translation system to translate the coded information into a complex functional protein.

How reasonable is it to imagine that purposeless random chemical interactions designed a DNA molecule capable of encoding information? How reasonable is it to assume unguided chance designed and imposed a highly specific, information rich coding system including necessary syntax and semantic elements into DNA? Is it reasonable to assume that, at the same time, unguided chance also designed a transmission and translation system that recognizes the DNA code and translates it into functional proteins and body parts? The information storage system, the coded information and the transmission and translation system must all be in place at the same time for functionality and viability.

I have provided a greatly simplified description of protein synthesis. I hope you get the idea that it involves elegant design and a complex orchestrated system consisting of many parts. You can see why the term *specified complexity* is an accurate description of biological molecules and systems. Living organisms are riddled with irreducibly complex systems such as this and I lack the faith that Darwinists have to attribute this to a Big Bang, and unguided time, chance and luck.

If you want a better and more detailed understanding of the process, I highly recommend you read *Darwin's Black Box: The Biochemical Challenge to Evolution* by Michael Behe Ph.D.

FOR FURTHER STUDY

CREATION WEBSITES

www.icr.org
www.answersingenesis.org
www.trueorigin.org

INTELLIGENT DESIGN WEBSITES

www.arn.org
www.discovery.org/id/

Darwin's Black Box: The Biochemical Challenge to Evolution, by Michael Behe, published by Free Press a Division of Simon and Schuster, Inc.

CHAPTER FIFTEEN
ENZYMES TO THE RESCUE

The chemicals in our body act and interact in accord with what we call the laws of nature. There is no evidence that these natural laws, operating with the properties of physics and chemistry can form the information-rich highly complex structures and systems we know as living cells or organisms. The so-called natural laws at work in biology, physics and astronomy are not innovative or creative, they are conservational and operational. Natural laws are running the show and when we study them, we can often see how things work but they don't explain how planets and people came to be in the first place.

When it comes to biology, natural laws and chemistry can neither create nor sustain life. Life is dependent on a variety of information-rich systems and molecules which operate within natural laws but also constrain and manipulate natural laws in order to sustain life. We could talk about necessary biological systems like the respiratory system, the circulatory system, the reproductive system, the digestive system, the immune system and many others but let's just look at a necessary molecule, usually a protein called an enzyme.

Enzymes are molecules (usually proteins) that act as catalysts to react with other molecules on contact – they make things happen. The reaction can build or break chemical bonds in order to combine smaller molecules into larger molecules or to break large molecules into smaller ones. At the end of the reaction, the enzyme is unchanged and can go on about its work. The newly created or modified molecules are necessary for the life and health of the cell or organism.

There is a lot of chemical activity going on in every cell and the average human has something around 40 trillion cells. We are dependent on things going right in our cellular makeup. If the molecules in our cells were left to unguided time and chance, they would interact and form some useful molecules but not quickly enough to sustain life. We are entirely dependent on thousands of enzymes, each designed to interact with a specific molecule to make the molecules we need with the speed and in the quantities we need to stay alive. Enzymes make things happen that random chance and the laws of nature could not accomplish in time to keep us alive.

Enzymes help us digest food, produce energy necessary for cellular life and many other functions necessary for life. Sometimes multiple enzymes act in concert in a specific sequence (called a pathway) to accomplish a necessary result. The recipes (genetic code) to create enzymes and all other proteins are found in our DNA.

To avoid getting too complicated I will only mention that the body also produces proteins that are enzyme blockers. These enzyme blockers can slow down or stop (regulate) certain enzyme activity and are part of our metabolic system.

Enzyme molecules are necessary for our survival and enzyme blocking molecules regulate enzyme activity. This is just one of many delicately balanced systems required for our survival. As we discuss an example of how enzymes work together in a complex choreographed way, ask yourself if this could be the unintended product of unguided time and chance. Is it apparent design or obvious design?

The more I study biological systems, the more I find my intuitive belief in creation is vindicated by science. Darwinists often accuse those who subscribe to creation/design of not understanding science. In reality, the more we understand science the more our creation/design position is affirmed. When Darwinists accuse creation/design advocates of not understanding science, they are referring to their version of science that has been conflated with naturalism/materialism. It escapes them that one can accept and evaluate the data of science without locking into the Darwinist materialistic worldview. The Darwinists are essentially saying, "If you don't believe what I believe, you can't be a real scientist." That would be news to Isaac Newton, Louis Pasteur and many thousands of real scientists in the past (see chapter four) and living today who subscribe to the creation/design model.

I've mentioned that enzymes are essential for our survival. To illustrate this, we will look at the amazing blood clotting mechanism and the role enzymes play in keeping us alive. The blood clotting mechanism is necessary for our survival, but it must be carefully regulated. When we need to stop the loss of blood, the clotting system must be in place and ready to respond immediately. At the same time, we can't afford to have uncontrolled clotting because if the flow of blood to a vital organ stopped, we could die. The blood clotting system must be in place and ready to spring to action at a moment's notice, but we can't have uncontrolled, freelance blood clotting activity or we would die.

When we receive an injury and vessels are broken, blood begins to flow. Our blood is under pressure to drive it to all parts of the body so when vessels are damaged blood will flow out of the vessel. We suffer many unnoticed vessel injuries every day due to bumps and friction from our daily activity. Serious injuries result in significant loss of blood and without a mechanism to stop the flow of blood, life is threatened.

I remind the reader to not get bogged down in the complexities of the blood clotting system. Just marvel at the incredible, irreducibly complex functional system that makes life possible and ask yourself if this level of design requires a designer or is it the natural unintended product of cosmic flatulence (a big bang followed by an expanding cloud of mostly hydrogen gas)?

Let's say you have an accident that cuts a vessel in your arm and blood begins to flow. The first thing that happens is chemical changes from the injury inform the smooth muscles around the vessel to contract to restrict blood flow. Isn't it fortunate that muscle cells are designed to react this way to certain chemical stimulation?

The blood throughout our body carries red blood cells that carry oxygen, white blood cells that fight infection, and platelets that play a role in blood clotting. Normally platelets are not sticky and travel through our blood in orderly fashion. However, when chemical changes at the site of an injury inform platelets, they immediately become sticky and stick to each other and the vessel walls to form a soft plug to help stem the loss of blood. Isn't it fortunate that mindless natural processes somehow anticipated the need for platelets and put them in our circulatory system so they are there when needed to help block blood loss in the event of injury? It's even more fortunate that there is a chemical signaling system at the site of an injury instructing the platelets what to do. Darwinists assure us there is no design here, so life is indebted to chance and a heap of luck.

The constriction of vessels and the soft platelet plug are important factors in stemming the flow of blood at the site of an injury but something more is needed to make a more substantial plug. What we need here is some sort of a net to hold everything together while the body goes about making repairs at the site of the injury. Of course the repair system is equally amazing but let's stick to stopping the flow of blood.

How lucky can we get? It turns out that the liver produces a linear molecule that, under the right conditions can be part of the net we need to strengthen the platelet plug and allow a stronger clot to form. We will think of this molecule, a string of amino acids, as potential netting. This potential netting material is called *fibrinogen* or clotting Factor I. So, the liver is pumping this potential net material into the blood but this potential netting molecule is not able to link end to end with other potential netting molecules to form a net because the linkage is blocked by chemical groups at both ends of the molecule. This is a good thing because if all the potential netting material flowing in our blood decided to freelance, link together and build netting in our circulatory system throughout the body we would die.

As luck would have it, when platelets are activated to form a platelet plug, they activate receptors that the potential net material (fibrinogen) attaches to. While potential net making molecules are collecting on the platelet plug, they are not yet able to form the strong net required. It's again quite fortunate that the liver is pumping this potential netting material into our blood stream, but it can't make a strong net without more help.

At this point, help comes in the form of an enzyme called *Thrombin*. You can call this enzyme "Mr. T" if you like because Mr. T comes along and cuts off the chemical groups at both ends of the potential netting material (the fibrinogen molecule) allowing the potential netting material to bond or link with other potential netting material molecules to make a strong fiber net. This new netting material molecule formed by activated fibrinogen is called *fibrin*. Fibrin is a strong fiber strand of linked fibrinogen molecules activated by Mr. T.. Mr. T also activates another chemical factor that allows the fibrin strands to link across each other as well as end to end making the strong netting required to keep the platelet plug in place and to allow necessary clotting. Can unguided chance anticipate the need for these molecules to be present in our blood and design a feedback system to activate them when needed?

At this point you may be wondering, where does Mr. T (thrombin) come from? If Mr. T is present in the blood all the time as a busy enzyme it would convert potential netting to actual netting (fibrinogen to fibrin) and we would all clot to death.

This isn't complicated enough yet but stay with me a little longer and prepare to be amazed. Mr. T (thrombin) isn't everywhere in our blood all the time or we would clot to death. But wait, Mr. T (thrombin) is in the blood but disguised in another molecular form called prothrombin (also known as clotting factor II). Under the right circumstances, another enzyme called *prothrombinase* forms and breaks certain chemical bonds in prothrombin to release Mr. T (thrombin). Once out of his molecular disguise (activated), Mr. T proceeds to convert the potential netting material into netting material (fibrinogen to fibrin) so a strong clot can build.

Now, you might be thinking that if the prothrombinase enzyme is always around in the blood, what would prevent it from converting prothrombin to thrombin which would convert fibrinogen to fibrin causing us to clot to death? That's a good question but no worries. The prothrombinase enzyme molecule is only created when needed at the site of an injury. Chemical "messages" at the site of an injury activate complicated pathways to form prothrombinase when and where needed.

In addition to more significant injuries from time to time, we suffer many unnoticed injuries that cause bleeding every day. Our blood clotting mechanism is at work every day to keep us alive. Our body (mostly in the liver) produces a variety of molecules, including enzymes and puts them into our blood stream to stop bleeding and repair injuries. These molecules must be in our circulatory system in the right amounts and balance at all times. The blood clotting system is a complex and coordinated system and strong evidence for design.

Our blood clotting mechanism consists of several molecules working together in a controlled way. Michael Behe PhD (molecular biologist) would refer to this as an *irreducibly complex* system. By that it is meant that every component of the blood clotting system must be in place for the whole system to work. If our body could not produce any one of the molecules mentioned in this discussion, life would not be possible. Our bodies are full of irreducibly complex systems, so this is an important concept to understand.

The Darwinian paradigm requires that all the components of an irreducibly complex system come together as a result of periodic mutational changes that provide some reproductive or survival benefit so they can be "selected" by natural selection. However, in the case of an irreducibly complex system, there is no survival benefit until all the components are in place. No single mutation has ever been observed to create a complex and highly orchestrated system like the blood clotting system. I don't imagine anyone, even a committed Darwinist, would claim that it could occur in a single mutational event. Yet the incremental development of a complex, information rich, coordinated, system where there is no survival advantage until all the individual components are in place is too much to ask of random mutations and natural selection. There is no advantage to "select" until all components are in place. Our bodies are not wasting energy and resources producing random functionless proteins and enzymes in hopes that a necessary beneficial irreducibly complex system might rattle into place.

The genetic code is similar to computer code in many ways. Is there a realistic expectation that having random code elements added or removed from a software program will produce a new and better program? Entropy (the second law of thermodynamics discussed in chapter 8) precludes it and our experience and intuition affirm it. We can name many diseases that are the result of mutations, but no irreducibly complex system has ever been observed to occur by genetic mutational accidents.

Bear in mind that the blood clotting system is just one of the many irreducibly complex systems that are necessary to sustain life. I hope the reader can appreciate the mind behind elegant, information rich, irreducibly complex biological systems that make life possible.

FOR FURTHER STUDY

CREATION WEBSITES

www.icr.org
www.answersingenesis.org
www.trueorigin.org

INTELLIGENT DESIGN WEBSITES

www.arn.org
www.discovery.org/id/

Beyond Irreducible Complexity: Natural Survival Capacity http://www.arn.org/bic.htm This link connects to a fascinating series of articles by Dr. Howard Glicksman who describes incredible evidence of design in many biological systems, including the enzyme system.

Darwin's Black Box: The Biochemical Challenge to Evolution, By Michael Behe, PhD published by Touchstone

CHAPTER SIXTEEN

THE HUMAN BRAIN

The human brain is an amazing organ with astounding capabilities. The more science discovers about the brain, the more incredible it is to believe that it is the unintended product of an expanding cloud of predominantly hydrogen gas after a distant Big Bang. The Darwinist attributes the functional complexity of the brain to time, chance, the natural properties of matter and a lot of fortunate mutations coupled with natural selection.

Many people are prone to accept the Darwinist position on the origin of life and evolutionary change, including the naturalistic development of the human brain, because they are convinced that science "knows" how it all happened. The reality is Darwinists don't "know" how it happened, but they "believe" it all happened as a result of unguided time and chance. Darwinian science is constrained by their belief in naturalistic/materialistic philosophy, so their conclusions are limited or dictated by their belief system (faith). While Darwinists in the public education system and popular media insist that we interpret scientific data and reality itself through their belief system, you are not constrained by their belief system. You are free to consider Darwinism's mechanistic chance or a Creator's design or any combination of the two.

There is much science and there are many scientists who disagree with the Darwinian model. For many of us, our everyday life experiences, logic and intuition cause us to reject the Darwinian model. When we look at the human brain as we have looked at other biological systems, I hope the reader weighs the Darwinian model based on unguided natural processes, time and chance against the creation/design model.

Most of the material I am presenting here was extracted from an article by a geneticist by the name of Jeffrey Tompkins Ph.D.. I found the article in *Acts and Facts*, a publication by the Institute For Creation Research – a link to an on-line version of the article is provided below.

Science has much to learn about the human brain but here's some of what we know so far:

1. A single synapse (the point where one neuron/brain cell passes signals to another neuron) is not a simple on/off switch as previously thought but is like a computer's microprocessor containing both memory storage and information processing features. One synapse can contain about 1000 molecular scale microprocessor units.

    An average human brain contains 200 <u>billion</u> nerve cells (neurons) connected to one another through hundreds of <u>trillions</u> of synapses.

    A single human brain has more information processing units than all the computers, routers and internet connections on earth. That's some amazing hardware but think about the firmware and software involved.

2. Dendrites are branch projections from brain cells that reach out to connect with other brain cells through synapses. Dendrites were thought to be passive conduits but have been discovered to generate 10 times more electrical spikes than the main body of brain cells. Dendrites are a hybrid system performing both digital and analog transactions like a quantum computer at warp speed levels.

3. The human brain memory storage capacity has been discovered to have a capacity of at least a petabyte (a petabyte is1000 terabytes. A terabyte is1000 gigabytes, a gigabyte is 1000 megabytes, and a megabyte is 1,000,000

bytes of information) …in the same ballpark as the World Wide Web. Isn't it amazing that the brain can store non-physical information (see chapter 12) in a physical matrix where it can be accessed as needed? Is it just me or does anybody else feel they are not putting all this capacity to good use?

4. The human brain requires a lot of energy and consumes about 20% of the body's energy budget. While the brain requires a lot of energy it is extremely efficient. It has been estimated that a man-made computer with a computing capacity of the human brain would require 10 megawatts to operate properly. This is comparable to the output of a small hydroelectric power plant. By comparison, the human brain requires only about 10 watts to function. How's that for efficiency?! Perhaps this explains how some people appear to blow a fuse!

5. Science still doesn't know a lot about how a brain processes information but it recently discovered that when a brain processes information, a thought or a task, it sets up a temporary interaction between a group of cells called a clique. Just as a group or clique of people can gather to discuss a subject, the brain assembles multi-dimensional structures of interactive cells to process a task and then the association disintegrates. The brain can have tens of millions of these temporary structures in a tiny area of the brain.

6. Science has recently discovered that not only does the brain use chemically generated electrical pulses to communicate, it uses photons (light particles) to communicate and process information. It appears that light channeled by filament structures helps coordinate activities in different regions of the brain. Since light travels faster than electricity this greatly speeds up processing. It looks like God developed fiber optics before we did. Unable to resist a pun, I can't help but point out that photons in the brain shed new light on our understanding of the brain.

Bear in mind that the brain doesn't mechanically store all visual, audio, mental input. The brain is dynamic and able to receive and store new information but also able to sort through input and "know" what is important to store for future reference and what is unimportant and can be discarded. This allows us to focus our thinking on what is important without the clutter and distraction of irrelevant input. Our amazing brain is able to receive, process and store information as well as, retrieve, process and output immaterial information. Let your brain process that. Is creation/design a reasonable consideration?

The human brain with its many billions of cells (neurons) and many trillions of synapse connections is the most complex organization of matter known to man. Research into the human brain continues and it has been found that there are significant differences between human and animal brains. Is the brain an unintended product of time and chance as required by Darwinism or is it the product of an amazing creator or designer?

FOR FURTHER INFORMATION

CREATION WEBSITES

www.icr.org
www.answersingenesis.org
www.trueorigin.org

INTELLIGENT DESIGN WEBSITES

www.arn.org
www.discovery.org/id/

The Human Brain is Beyond Belief, article by Jeffrey Tompkins Ph.D. available at: https://www.icr.org/article/human-brain-beyond-belief

CHAPTER SEVENTEEN
WHAT ABOUT BIOLOGICAL CHANGE?

How does the creation model explain the obvious evidence of change we see in animals living today and in the fossil record? This is a good question because fossils and biological change are usually presented as evidence for Darwinian evolution. Change in the Darwinian model is based on time and chance or, more specifically, mutational change and natural selection. In this chapter, I will discuss the Darwinian model but then go on to explain biological change from the perspective of a Biblical creationist. Creationists don't accept Darwinian evolution as described in chapter two and refined in chapter four, but that doesn't mean we don't recognize and account for change in living organisms.

In previous chapters I have pointed to some of the flaws inherent in Darwinism. Founded in naturalism/materialism, Darwinism relies on the properties inherent in matter and a throw of dice, in the form of genetic mutations, to produce changes that drive biological evolution. These changes are mechanistically "selected" through a process called "natural selection" to advance living organisms in the struggle for adaptation and survival.

I have tried to make the point that the case for creation/design is not based solely on the flaws in or inability of the mechanisms of Darwinism to account for the origin and diversity of life. Life is dependent on design and information and information only comes from a mind, an intelligent source. Life involves complex information storage and processing systems and mechanisms which anticipate and respond in both mental and physical ways to environmental stimulus.

A problem with Darwinism's mutation/selection model is that science has determined that living organisms exhibit incredibly complex design orchestrated by highly precise coded information. Actually, we are seeing multiple interacting coded information systems in living organisms. Information science informs us that information doesn't originate in matter. It is imposed on matter by an intelligent agent. I've tried to illustrate this with the statistical impossibility of chance to produce complex, highly specific information rich biological molecules. It is unreasonable to expect random mutational changes to genetic code, like random changes to a computer software application, to bring about an improved result. In addition, physical matter has no inherent selection capacity to recognize beneficial mutational changes should they occur. Naturalism and materialism actually preclude the involvement of an intelligent selector to choose supposed beneficial mutations. To account for the massive amount of information and obvious evidence of design in living organisms, Darwinists often refer to "apparent design" and personify nature so that Mother Nature can be the selector that finds a way. In Darwinian theology, nature has no purpose, so Mother Nature personifies purposelessness to give natural selection the ring of credibility where random chance has no hope of producing information rich biological organisms. Darwinism's apparent design requires an apparent designer, so Mother Nature fills the bill.

My goal here is to help the reader to see and think outside of the naturalistic/materialistic box imposed by Darwinism. Design requires a designer and information requires an intelligent source. We "know" this by experience and intuition and it is illustrated by science when unrestricted by naturalistic/materialistic philosophical presuppositions. Many of the changes we see in diverse living organisms are programmed responses to allow adaptation to new or changing environmental conditions. The selecting agent in biological change is not a personified Mother Nature, it's the organism itself. More precisely, the change options were pre-selected by the creator/ designer and coded into the organism's genetic software to enable it to respond to the environment in particular ways. Science continues to uncover some of the ways organisms are programmed to respond and adapt to the environment. Remember, programming (including genetic programming) is information and information is immaterial but is imposed on matter by an intelligent agent.

It is increasingly apparent that many organisms are pre-programmed to adapt to changing environmental conditions. Scientists at the *Institute for Creation Research* are studying what they call the CET model. The CET stands for Continuous Environmental Tracking where the organism draws on what are essentially pre-programmed adaptive responses to various environmental conditions. Some of these adaptive response options are found in the genetic and epigenetic information systems. It appears the missing selector personified by Mother Nature in the Darwinism model is found in the organism itself. It will be interesting to see where this CET research leads.

The intelligent agent responsible for life and biodiversity isn't Mother Nature but we will explore that in chapter 19. For now, let's look outside the time and chance, mutation/selection box of Darwinism and consider biodiversity from the perspective of a creationist.

In the creation model, as in the Big Bang model, the universe had a beginning. The beginning was not a Big Bang that popped everything into existence out of nothing, but an eternal omnipotent God who created everything we know (our universe) ex nihilo (out of nothing). Our creator God transcends and is not constrained by the natural laws that are part of his universe, he created them in an act of special creation. The universe now operates by the natural laws that are part of the super-natural creation. In essence, what we see about us and refer to as the natural world, is in reality a supernatural world that time and chance and natural laws could never create. The unique properties of particle physics are finely tuned to allow matter to exist and, under the right conditions, for living organisms to exist and thrive.

If we look with an open mind, we can see God's fingerprints all over the creation. As I mentioned in earlier chapters, I believed in creation and a creator before I knew who the creator was and before I studied the science that supports creation. I "knew" it by experience and intuition. I don't think that makes me unique.

In the process of creation, God created the earth with the unique properties required to support life. There is much that could be said about the special conditions on earth, distance from the sun, presence of liquid water, breathable air, the moon, our axis tilt, etc., but my point is that the earth is "special" in the universe.

God created many different kinds of plant and animal life equipped to live in many different environments. As the Bible puts it in Genesis, God created life as different "kinds" and each kind was intended/instructed to reproduce and "bring forth after their kind." Man*kind* was created uniquely different from all other animals, in the image of God.

God intended the various kinds of life he created to multiply and fill the earth. To enable them to do this, he gave the created kinds significant genetic diversity to enable organisms to adapt to different and changing environmental conditions. For example, during times of drought some plants can adapt by growing deeper roots or accessing other resources already in their genetics. The finches Darwin observed on the Galapagos Island had thick beaks and differed from finches he saw on the mainland. It turns out that years of drought had changed the food source on the islands to hard seeds and the thick beaks were beneficial under these conditions. After several years of good rain, the food source changed, and the thick beaks were no longer necessary and went away. Darwinists cited this as an example of natural selection driving evolutionary change. However the finch bird has the genetic information for a thick or thin beak and there is no selector external to the bird so the change or adaptation was driven by a mechanism within the bird, likely the environment sensing epigenetic controls described in chapter thirteen.

When organisms change by drawing on their genetic diversity to adapt to changing environmental conditions it seems more accurate to think in terms of the organism responding to the environment rather than the environment and natural selection acting on the organism. In any event, we have noted that the environment has no selecting mechanism while the organism does. Adaptation by sensing environmental changes and selecting from preexisting genetic resources is much more effective than waiting and hoping for lucky mutations to save the day.

We get a good idea of the dramatic genetic diversity God has given organisms when we look at how humans exploit this genetic diversity by artificial selection (as opposed to natural selection). We exploit preexisting genetic diversity to develop better smelling roses, more protein in a grain, cows that give more milk or less cream, tomatoes that have a longer shelf life for the market, etc.. We should not equate a created "kind" with what we call a species because a number of species can potentially be derived from a single created kind.

Look at what we have done with the dog kind. Even Darwinists believe all dogs originated from an original wolf-like dog. We humans have selected diverse genetic traits within the dog kind to "create" a wide variety of dogs in different shapes and sizes with different traits to suit our fancy. As we artificially select and breed plants and animals for different traits, we thin the genetic diversity in the various varieties we create. These selected varieties are not as hearty as the original created kind. For example, a chihuahua or basset hound won't do as well in an artic climate as a wolf or husky, but we like our chihuahuas and bassets. Like our pets and other animal varieties we have created through artificial selection (selective breeding), we have created many different flower, vegetable, fruit and grain varieties. However, most of the varieties we have created are not as hardy as their wild cousins and require special conditions and care for survival.

Genesis tells us that God created plants and animals that reproduce after their kind. This means dogs produce dogs and cats produce cats. When birds reproduce, they produce birds not frogs or bats. When we look at animals that are interfertile, even if they don't normally reproduce in the wild, we can know that they are derived from an original created kind that reproduces after its kind. The fact that horses, zebras and donkeys are interfertile tells us they are derived from an original source kind. Lions and tigers don't usually mate together and reproduce in the wild, but they are interfertile so derived from an original kind. Sometimes it is clear that animals likely from an original kind have lost the ability to be interfertile. For example, a mule is a cross between a horse and a donkey, but mules are usually sterile. They are clearly descended from an original created kind but in the genetic downstream, fertility can sometimes be lost or limited. Genetic lines show unexpected and interesting relationships. House cats can interbreed with some small wildcats and some ocelots can breed with some house cats. Leopards and tigers can be interfertile and leopards and pumas can be interfertile and pumas have interbred with some ocelots.

When we look at the fossil record, we see a lot of extinction, but we see cats came from cats, dogs from dogs, bats from bats, etc. We see change within kinds, but we don't see all the transitional forms that must exist between all the different plants and animals if Darwinism is true. We do see changes in many cases, but we don't see evolution in the sense that one type of organism has evolved into another type. We do see new species arise in the fossil record, but a species is generally defined as a unique interbreeding population. Lions and tigers are different species but interfertile and thus derived from an original created kind.

Darwinism often portrays a graphic called the phylogenetic tree sometimes called the tree of life. At the base of the tree is a single-celled organism representing the first life to have evolved. As the tree trunk rises there are different branches coming off the trunk representing the evolutionary origin of different plant and animal life all derived from the original living organism at the base of the "tree." Somewhere near the top of the tree there is a primate branch with a human at the end of this branch. This phylogenetic tree diagram has explanatory value for a Darwinist, but it doesn't represent what we see in the fossil record. It is evolutionary art representing what Darwinists believe, not what we actually see in the fossil record.

If every living plant and animal evolved from an original single celled organism the fossil record should be characterized by (riddled with) transitional fossils as one type of organism gradually evolves over thousands or millions of years into a completely different organism. The millions of transitional forms that lived and died should be reflected in the fossil record. However, the fossil record reflects the sudden appearance of fully functional distinct organisms. While most people think the fossil record is evidence or proof of evolution, the reality is that the fossil record is characterized by the systematic absence of transitional organism. The fossil record supports creation more than evolution. We see there is some change over time, but in many cases, there is extinction, or the living organisms today are identical or nearly identical to the earliest fossil example.

The systematic absence of transitional fossils is not something that is only noticed by creationists. Charles Darwin noticed it and the late paleontologist Stephen J. Gould, an evolutionary biologist at Harvard and an outspoken Darwinist, said, *"The extreme rarity of transitional forms in the fossil record persists as the trade secret of paleontology. The evolutionary trees that adorn our textbooks have data only at the tips and nodes of the branches; the rest is inference, however reasonable, not the evidence of fossils.*"[1] Gould also said, *"The history of most fossil species include two features particularly inconsistent with gradualism: 1) Stasis - most species exhibit no directional change during their tenure on earth. They appear in the fossil record looking much the same as when they disappear; morphological change is usually limited and directionless; 2) Sudden appearance - in any local area, a species does not arise gradually by the steady transformation of its ancestors; it appears all at once and 'fully formed'.*"[2]

To account for the missing transitional forms in the fossil record, Gould promoted the theory of *punctuated equilibrium* described below.

Life on earth is better represented graphically as a forest rather than a single tree. Each distinct kind of organism (each tree in the forest) represents an original created kind with some variation (branching) but not connecting to another tree (kind). The single phylogenetic tree diagram is the only model consistent with Darwinian evolutionary presuppositions, but it doesn't reflect what we see in the fossil record and living today. Darwinism requires transitional fossils, so they do cite some changes in fossil organisms as transitional to another type of organism but often these claims are highly disputed even within Darwinian circles and the changes are often just natural variations. If Darwinism is true, the fossil record should be filled with obvious transitional forms as all the varieties of life evolved from an original single-celled organism. Darwinists would not be reduced to arguing over the occasional supposed transitional forms.

The systematic absence of transitional fossils is so prevalent in the fossil record that Darwinists proposed the concept of *punctuated equilibrium* to account for the missing transitional fossils. The idea of punctuated equilibrium is that organisms tend to remain stable (in equilibrium) until there is a rapid drastic change in a small isolated population that produces a significantly different variety which out competes and replaces the original population. This sudden rapid change conveniently happens in a small isolated population to account for the missing fossils. There is no known and observed biological process to account for the sudden rapid change but that's not a serious problem if you are to save Darwinism.

Punctuated Equilibrium was fairly popular for a while but since it was based on missing fossils and unknown biological processes it appears to have lost popularity. I think it was too close to a previously proposed and rejected *hopeful monster* proposal where sudden major biological changes happen all at once in a single generation; like a lizard lays an egg and out comes the first chicken. The hopeful monster and punctuated equilibrium concepts were intended to explain the missing transitional fossils but being based on missing fossils and no known or observed biological processes I suspect they were too close to the miraculous to survive in the Darwinian paradigm. Darwinism appears to have retreated to the gradual mutation/selection model of change which ignores the sudden appearance of organisms and missing transitional fossils.

FOR FURTHER STUDY

CREATION WEBSITES

www.icr.org
www.answersingenesis.org
www.trueorigin.org

INTELLIGENT DESIGN WEBSITES

www.arn.org
www.discovery.org/id/

1. Gould, S.J. (1977) Evolutions Erratic Pace, Natural History. Vol. 86 (5): 12-16

2. Gould, S.J. (1977) Evolution's Erratic Pace, Natural History, Vol. 86

CHAPTER EIGHTEEN
CLOSING THOUGHTS AND SUMMARY

The public education system in the United States and elsewhere has conflated the discipline of science with naturalistic/materialistic philosophy. Put another way, the discipline of science has been defined within the constraints of a naturalistic or materialistic worldview or belief system. This philosophical position is, for the most part, reflected in public education and the popular media and has gained significant cultural authority.

The conflation of philosophical naturalism/materialism with the discipline of science is illustrated by a quote from Richard Lewontin, an American evolutionary biologist, mathematician, and geneticist:

"We take the side of science in spite of the patent absurdity of some of its constructs, in spite of its failure to fulfill many of its extravagant promises of health and life, in spite of the tolerance of the scientific community for unsubstantiated just-so stories, because we have a prior commitment, a commitment to materialism. It is not that the methods and institutions of science somehow compel us to accept a material explanation of the phenomenal world, but, on the contrary, that we are forced by our a priori adherence to material causes to create an apparatus of investigation and a set of concepts that produce material explanations, no matter how counter-intuitive, no matter how mystifying to the uninitiated. Moreover, that materialism is an absolute, for we cannot allow a Divine Foot in the door."

(Richartd Lewontin, Billions and billions of demons, *The New York Review*, p.31, January 9, 1997)

I have seen similar statements by other Darwinists. It seems that some Darwinists, having fully invested their faith in philosophical naturalism/materialism, are compelled to conflate the discipline of science with their faith. Often, those who hold this faith attribute to themselves some sort of intellectual superiority over those who propose a creator/designer. Regardless of the evidence, the committed Darwinist claims and clings to the assumed intellectual superiority of their naturalistic/materialistic worldview. While they think they are saying, "My science trumps your religion" they are actually saying, "My religion trumps your religion and your science."

With this philosophical predisposition, when it comes to origins science, naturalistic evolution is the only model possible. Any model of origins that includes concepts of creation or intelligent design is rejected out of hand before the data is even considered because these models are philosophically unacceptable.

Many people, like myself, have not been able to accept a purely naturalistic evolutionary model of origins because it's not consistent with our observations and life experiences. We don't see time, chance and chemistry organizing and creating the immense complexity and associated information we see in nature and living organisms. We humans fall in love, write poetry, make music and create art and we don't see this as a natural property of chemicals. Our intuition and logic tell us that our existence and life here on earth seems to be more than an unintended mix of chance and chemistry.

Just as non-physical information rides on physical elements, our life, personality, or spirit rides on, yet transcends, our physical self. The real us isn't physical, has no mass and is therefore not bound by physics and time. After all, time is a product of the physical universe and is controlled by such things as mass and velocity. Our information-reliant physical life is constrained by physics and chemistry which, in turn, are constrained by the second law of thermodynamics. Accordingly, our physical bodies are moving relentlessly toward ultimate entropy and death. However, our non-physical self, our spirit, while constrained by our physical body, is not bound by physics, chemistry and entropy. Just as information is immaterial but able to be expressed in material form, our spirit is immaterial, yet we see it expressed in material bodies. Just as information always comes from an intelligent source, our spirit came from our Creator God, the ultimate intelligent source. When our physical body obeys the laws of physics in death, our non-physical eternal spirit continues. In the next chapter we will consider eternity as we confront the creator.

In preceding chapters I have described the conflation of science with naturalistic/materialistic philosophy and how this philosophical constraint makes naturalistic evolution the only allowable model of origins and evolution. I have referred to origins science constrained by naturalism/materialism as Darwinism. Any model of origins or evolution that employs any element of creation or design is philosophically incompatible with Darwinism and is rejected in the public education system and, for the most part, in the popular media. While this rejection is said to be on the basis of science, it is clearly on the basis of a philosophical predisposition.

Preceding chapters have also outlined how unguided natural laws and the natural properties of matter and energy are unable to produce the information content and specified complexity of biological molecules and irreducibly complex biological systems required for life. The statistical probabilities of the chance origin of information-rich complex biological molecules like proteins and DNA, not to mention life itself, are beyond possibility given the age of the universe, even if extended many fold.

Nevertheless, while scientific data and statistical probabilities rally against Darwinism, the naturalistic origin and evolution of life is declared to be the only possible "scientific" model of origins and evolution. It's the only model taught in public education and receiving governmental support. Faith in Darwinism is so strong that we often hear people of science claim that our universe is so vast that life must certainly have evolved elsewhere as well as on earth. We see this reflected in science fiction films like Star Trek and Star Wars.

Based on Darwinian faith that life evolved by natural processes on earth and it likely evolved elsewhere in the universe, the SETI organization was established. SETI stands for Search for Extra Terrestrial Intelligence and the organization has existed in various forms for years receiving government assistance and various private funding. Today it is known as the SETI Institute and its mission statement is, "to explore, understand and explain the origin and nature of life in the universe and the evolution of intelligence."

The SETI organization has been scanning our noisy universe for years with radio telescopes trying to detect signs of intelligence in the noise of the stars. There is every confidence that they will be able to distinguish information or intelligence from the random natural radio signal noise of the universe. This ability to detect information or intelligence was illustrated in the movie Contact. In essence, they are looking for information revealing an intelligent source in the noise of the universe.

I find it beyond ironic that scientists are scanning the sky with radio telescopes for information revealing extra-terrestrial alien intelligence yet when they look at biological molecules through microscopes and see vast amounts of information, they fail to see an intelligent source. This is how Darwinism works; information from outer space would reveal alien intelligence but information in biology reveals natural processes. Of course, Darwinism would maintain that aliens "evolved" by natural processes, but the point is they know information only comes from an intelligent source…. except in living organisms where it comes by time and chance.

As a young person I found myself in somewhat of a bind. I had an intuitive belief in creation, but I was told science explains everything by natural processes. I believed there was a God who was involved in the creation, but my family was not religious, and for a long time, I had no definitive concept of God. I had great respect for the discipline of science, but I could not entirely invest in the purely naturalistic model of origins presented to me. I believed in creation/design, but I was led to believe that science had proved evolution to be true. Even after I became a Christian in late high school and through college and beyond, I was conflicted by the creation/evolution issue.

Many years beyond college I began to read books by scientists addressing the creation/evolution issue. As I did my homework, I saw that there is powerful scientific evidence for creation or, if you prefer, intelligent design as opposed to naturalistic evolution. I also learned that the discipline of science need not be constrained by the philosophies of naturalism or materialism. These philosophies are not science, they are worldviews that some people bring to, impose on, or use to constrain science. This is especially true when we investigate origins science. If science is to be objective, one should be able to consider and follow all the data without philosophical constraints.

My objective in this work is to encourage and validate those inclined to believe in creation or design. Darwinists (those who conflate science with naturalism/materialism) do not know something you don't know about science; they believe something you don't believe or don't have to believe. The real battle is worldview vs worldview not religion vs science. I hope I have inspired or encouraged others to do their own homework in the origins debate. Learn to differentiate science from philosophy and follow the data to the best conclusion rather than just the conclusions allowed by the constraints of Darwinism.

In this work I have occasionally poked fun at Darwinism. My intent has not been to ridicule anyone but to provoke thought. I recognize that there is a lot of heat and smoke generated in the creation/evolution debate. Advocates of creation or intelligent design are often attacked and ridiculed by Darwinists. I have not approached the level of vitriol I sometimes see directed at creation/design in Darwinist literature.

While Darwinists love to portray the creation/evolution debate as religion vs science, I have made the point that it is not religion vs science but a worldview vs worldview matter. The above quote from Richard Lewontin illustrates this point. Any worldview that incorporates creation or design is entirely at odds with a naturalistic/materialistic origins model.

We could argue about whether science ever *proves* anything, but for the sake of the argument I will point out that if creation or design were proven to be true *in any capacity*, it would utterly falsify Darwinism which restricts the science of origins to an entirely naturalistic/materialistic model as a matter of faith. Creation/design destroys the religion of Darwinism and therein is the source of the heat and smoke. Darwinism defines science within the constraints of the naturalistic/materialistic worldview in order to preclude science from destroying their underlying worldview. Darwinism seeks to win the origins debate by definition rather than submitting their model of origins to public and academic scrutiny beside creation/design models.

FOR FURTHER STUDY

CREATION WEBSITES

www.icr.org
www.answersingenesis.org
www.trueorigin.org

INTELLIGENT DESIGN WEBSITES

www.arn.org
www.discovery.org/id/

CHAPTER NINETEEN
CONFRONTING THE CREATOR/DESIGNER

Darwinists often accuse creationists of believing in a god of the gaps. By this they mean that when science is unable to explain something, creationists insert their God into that gap in science and claim, "God did it." Hopefully, I have demonstrated that creationists are not calling on a God of the gaps in science. It's not what we don't know about science that makes us creationists, it's what we do know about science that convinces us of creation/design. A Creator God is not suggested by the gaps in science but by the whole of science as well as our experience and intuition.

Whether you subscribe to a creation or intelligent design model of origins, there necessarily must be an entity, a creator or designer, who acts outside of and is unconstrained by natural laws, time, and chance. I suppose you could argue that God used natural laws to create everything but then you are still stuck with describing the origin of natural laws, matter and energy and how unguided natural laws could create everything including complex living organisms. I have pointed out the incapability and statistical impossibility of unguided matter creating information and living organisms. You can't logically argue that everything happened naturally, and God did it. If God did it, then it's not natural, it's supernatural and philosophical naturalism, materialism and Darwinism are falsified.

The deist philosophy holds that there is sufficient evidence to suppose a supernatural entity or god exists and created everything but this god doesn't intervene in his creation, i.e., the universe. This position embraces naturalism in the present but is incompatible with Darwinism because it asserts a supernatural origins model. The god of deism cannot be the God of the Bible who clearly intervenes in the lives of individuals and cultures. The God of the Bible also acts within the physical universe with what we call miracles as recorded in the Bible and as experienced through the ages. However, the God of the Bible doesn't ordinarily intervene in the operation of the physical universe. Today it operates in conformance with what we call natural laws which He created supernaturally.

Many people are uncomfortable with the idea of a God or intelligent designer even though they are convinced of design in nature. I suspect that some people refuse to identify the creator/designer because once they do that, they can be put in a "religious" box. They are likely, in our current cultural environment, to be labeled as a religious nut or too religious to understand science or to function as an objective scientist. In our culture, belief in creator/design is considered a bias while belief in materialism or naturalism is considered to be objective. We've dealt with this in chapters three and four.

The Darwinist who believes in a god must put that god in a box that prevents him from being the creator. Nevertheless, I challenge the reader to do your homework and confront your creator/designer. Don't put him in a box of your making. After all, if there is a creator/designer, we are living within his box, that is, within the context of his creation and purposes. In this case, it would do us well to know and honor him on *his* terms.

I have often pointed out that the creation vs evolution debate is, at its core, a clash of worldviews. There is scientific data to be considered but often the conclusions reached are filtered through and sometimes constrained by philosophical predispositions. The Darwinist's naturalistic/materialistic worldview screens out any possibility of creation or design to conclude we human beings are here on this planet by chance outside of any grand plan or purpose. This isn't a decision based on evidence, it's a choice.

Human life, in the Darwinist view, is an unintended outcome in which a chance combination of chemicals is able to experience a temporary flash of consciousness in a vast meaningless space time continuum. The Darwinist who subscribes to naturalism as opposed to materialism may believe in a god, but they must put that god in a box because he can't be a real designer or creator. True Darwinism rejects purpose, design and creation because evolution is, by their definition, a natural, not a supernatural process. The Darwinist's god is limited to a spiritual domain as he can't alter, tinker or interfere in the physical domain where only natural things happen. The Darwinist who claims god used evolution to create (as I once did) must confront and explain the missing transitional fossils and the scientific and statistical obstacles we have discussed. Alternatively, they must claim that their god monkeyed with chemistry and mutational changes to produce life as we know it. However, this option is not available to a true Darwinist because Darwinism is based on natural processes alone and does not accommodate a god who monkeys with physics, chemistry or biology.

A non-creator god is essentially unnecessary, and I struggle to see where such a god obtains authority to tell naturally evolved humans how to live their lives. After all, we would be here by natural processes whether the non-creator god exists or not. The Darwinist who claims *God used evolution to create* is falsifying the naturalistic/materialistic underpinnings of Darwinism. If God is allowed to create anything or nudge evolution along in any direction at any point, Darwinism is falsified, and that God can never be put back in a box. That is precisely the point Lewontin was making in the quote in the previous chapter when he said, "...*materialism is an absolute, for we cannot allow a Divine Foot in the door.*"

On the other hand, if we humans are the product of creation or design, then we exist within the context and purposes of our creator/designer. The idea of being accountable to a creator sends some scurrying to the (temporary) safety of Darwinism while others try to put the Creator God in a box such as cosmopsychism or panpsychism where they can keep him at a safe distance if not manage him.

There is much evidence for creation and design beyond what I describe in this admittedly shallow work. This evidence causes some to reject pure materialism in recognition of the reality and necessity of a mind behind the obvious evidence of creation/design. Cosmopsychism proposes a mind or intelligence precedes and is the causal agent of the physical universe but is independent of physical reality. Panpsychism proposes that the mind or intelligence is an intrinsic part of the physical universe. In any event, reducing a creator/designer to an impersonal "mind" (may the force be with you) puts him in a convenient box that avoids accountability to a personal creator. It also dodges messy theological doctrines and, with any luck, avoids one from being labeled as "religious" in any traditional sense. The idea that we exist on purpose within the greater purposes of a creator has significant implications that some prefer to avoid.

Within the Darwinist worldview there either is no god (materialism) or if there is a god, he is not a creator (naturalism) and, as such, has no authority to tell naturally evolved humans how to live. Within this context, every individual human being and culture can make up their own rules of right and wrong, good and bad, moral and immoral. There are no absolute standards by which to claim any act or morality is right or wrong.

On the other hand, in the creationist or design worldview, there is an absolute life-giving authority who has the right as creator to establish right and wrong, moral absolutes and accountability, within his creation. We are living in *his* universe, *his* box. Accordingly, there is an absolute moral authority by which to judge the lives and conduct of created human beings. There is a moral authority that transcends individual and cultural moral values.

The God of the Bible is not an impersonal mind or force; he is a personal God who declares himself to be the Creator of the heavens and earth and life itself. The God of the Bible, our creator, tells us who he is and who we are. I see a bit of irony when people define themselves and their creator differently to make him more manageable. God has revealed himself to us in his creation and we have been describing this in previous chapters as evidence of creation or design in nature. God holds us accountable for recognizing him in the creation. In Romans 1:20 we are told, "*For the invisible things of him from the creation of the world are clearly seen, being understood by the things that are made, even his eternal power and Godhead; so that they are without excuse.*" While Romans says one is "*without excuse*" for failing to see God in the creation, Psalm 14:1 is a little more blunt when it says, "*The fool has said in his heart, There is no God.*"

God has also put something into our hearts that tells us there's something more to life, something beyond life on earth. This is described in Ecclesiastes 3:11 *"He has made everything beautiful in its time. Also, he has put eternity into man's heart, yet so that he cannot find out what God has done from the beginning to the end."* We can see in creation, or what is called general revelation, that God exists. We know there is something more, but we can't quite grasp it in general revelation alone.

God has revealed himself and his purposes for mankind more fully in his Word, the Bible, and in person when he walked among us as the God-man Jesus Christ. So, we can know God exists from *general revelation* (the creation) but we can know him and his purposes and plans for us personally in his *special revelation*, the Bible and Christ.

The God of the Bible claims to be the creator and has affirmed it through miracles and Scripture which is the very Word of God. I don't expect every reader to accept that the Bible is the Word of God just because I say so. I encourage the reader to do their own homework, read the Bible and make inquiry. I have provided a number of helpful resources (links and books) below to help in your inquiry. I particularly recommend the book, *Evidence That Demands a Verdict* for anyone investigating the authority and reliability of the Bible as the Word of God.

The creation of everything (heavens and earth), the earth and life on it is described in the first chapter of the book of Genesis. The Genesis record was written in historical narrative style and not as poetry or allegory. It's not a scientific description but it is intended to be received as a literal account. God didn't use evolution to create, he created everything ex nihilo (out of nothing). Further details are found in the second chapter of Genesis and here and there throughout the Bible. For a scientific and theological analysis of the creation event I refer the reader to the creation websites listed below and particularly to the Henry Morris book titled *The Genesis Record.*

In Genesis we learn that the creator of the universe created mankind (male and female) in his image. Being created in the image of God has many implications but at the very least it is clear that we are set apart from all other animals with a mind and spirit able to communicate and have a relationship with God. Mankind is unique in many ways such as in our creativity and ability to communicate and contemplate our existence within the cosmos and eternity. Our ability to recognize and respond to right and wrong (moral) matters and the concept of justice is also unique to mankind.

God created us to have fellowship with him and to procreate, fill the earth and have dominion over his earthly creation. Along with this is the implied responsibility to manage and care for the initial perfect creation God provided for us. The first humans were Adam and Eve, two literal people from whom all of us are descended. God did not create different races; he created the human race in two people with the genetic diversity to account for what we call different races today. We are all one race before God and there is no such thing as a superior or inferior race of people.

As our creator, God established certain moral rules by which we should live. God created us as free moral agents so we can choose or reject to live by God's rules. From the very first couple to all of us today, we have all failed to fully obey our creator. This fault is what the Bible calls sin and it has separated us from the relationship we were intended to have with God. The word sin is derived from a Greek word used in archery which means, "to miss the mark."

Sin is also the reason there is death in the world today. In God's initial creation there was to be no suffering and death but that perfect creation fell into decay when sin entered the world. Today we live in a world that is beautiful in many respects but also full of disasters, suffering and death. Today we will all die because we all fall short of God's standards, we have missed the mark, it's our nature, we are all sinners before God.

We see the effects of sin in individuals, cultures and nations. We see greed, hate, envy, murder, sexual depravity, theft, wars and much more. It is plain to see that mankind is less than perfect. There is clearly a problem and the Bible calls it sin.

That we are all sinners is clearly stated in Romans 3:23 which tells us, "*all have sinned and fall short of the glory of God.*" That we will die because of our sin is stated in Romans 6:23 which says, "*...the wages of sin is death, but the free gift of God is eternal life in Christ Jesus our Lord.*"

Indeed, we see today that everybody dies but it was not intended to be that way in the initial Creation before sin.

The bad news is that we are all sinners and will die but God did not stop loving us because we are sinners. Our intended close relationship with our creator has been broken by our sin but the good news (the word Gospel means good news) is that God has provided a remedy. We get a hint at that remedy in the last half of Romans 6:23 where it says, *"but the free gift of God is eternal life in Christ Jesus our Lord."*

God is holy and righteous beyond our ability to fully comprehend. Heaven and Hell are described as real places in Scripture. God, being both righteous and just, cannot allow sin to go unpunished and so death is the penalty for mankind because we are all sinners. However, in spite of our sin nature, God loves us and has made a path for forgiveness and acceptance into his holy presence. This is expressed in John 3:16 where we are told, *"For God so loved the world, that he gave his only begotten Son, that whoever believes in him should not perish, but have everlasting life."* God's standard to be received into his eternal presence is not about who is better than who. God's standard is absolute perfection, and that perfection is found only in our Savior, Jesus Christ.

God *"gave his only begotten Son,"* what does that mean? It means that while God hates sin, he loves sinners enough to send his Son, Jesus Christ, into the world to die for sinners. As Romans 5:8 puts it, *"But God commends his love toward us, in that, while we were yet sinners, Christ died for us."* The righteousness of God requires that sin be punished by death but God himself, in the person of Christ, paid that penalty on the cross. Jesus Christ, God the son, became a man and lived a sinless life on earth. He was and is the only sinless person qualified to die for the sins of others because he had no sin of his own.

In Christ's death on the cross, the penalty for sin was paid on behalf of sinners but God doesn't impose this salvation on anyone. Forgiveness of our sins and being put right with our creator must be received by faith. A key concept here is faith. We must not only believe in God, but we must seek and believe God. As Hebrews 11:6 puts it, *"But without faith it is impossible to please him: for he that comes to God must believe that he is, and that he is a rewarder of them that diligently seek him."*

Our salvation is not something we earn, it's something we receive by faith. As Ephesians 2:8-9 puts it *"For by grace you have been saved through faith. And this is not your own doing; it is the gift of God, not a result of works, so that no one may boast."*

So, salvation is offered by God's grace, not earned but received by faith. But what are we saved *from* and *to*? Remember John 3:16 said, "… *whoever believes in him should not perish, but have everlasting life."* There may be death on earth but everlasting life with God in Heaven is offered through God's grace and our faith. Hebrews 9:27 says, *"And just as it is appointed for man to die once, and after that comes judgment."* Our salvation is *from* God's judgement for our sin and *to* eternal life with him in Heaven. We are eternal beings and death and judgement are physical and spiritual realities. Our eternity will be in Heaven or Hell so choose wisely.

Mankind's salvation and relationship with his creator God is possible only by God's grace and our faith. No wonder Christians sing of *Amazing Grace.* Normally grace or mercy is granted at the expense of justice, but not so with Christianity. In Christianity, grace and mercy are granted with justice because Jesus died on the cross to pay the price, the wages, of sin for all who will believe and receive Him.

Faith, believing God our creator, and Jesus our Savior is the basis of our salvation, the difference between eternity in Heaven or Hell. What must one believe?

Believe there is a God, your creator, he has given you all the evidence you need: Romans 1:20

Believe that God is perfect, but you aren't, you have missed the mark, you are a sinner: Romans 3:23

Believe that your sin condemns you to death: Romans 6:23

Believe that you must be sorry for, confess and turn away (repent) from sin in your life: Acts 3:19 *"Repent therefore, and turn back, that your sins may be blotted out,"* 1 John1:9 *"If we confess our sins, he is faithful and **just** to forgive us our sins and to cleanse us from all unrighteousness."* Remember, God's grace and mercy is not at the expense of justice. God is *just* to forgive sin because the price has been paid in full.

Believe that Jesus Christ died for and paid the penalty for *your* sins and receive him into your life as your Lord and Savior to have a personal relationship with your creator on earth and in eternity with him in Heaven. John 3:16, Romans 5:8, Romans 8:1 *"There is therefore now no condemnation for those who are in Christ Jesus."*

God's salvation is not imposed on anyone against their will. Salvation is available to all but on our creator's terms. One is not granted salvation because they believe in God, even the Devil believes that. James 2:19 says *"Do you believe that there is only one God? Good! The demons also believe—and tremble with fear."* God's salvation is a result of believing in God and believing God when he tells us who he is and who we are. Salvation is being *in* Christ Jesus, receiving his grace by faith. Saving faith is not something we say or do, it is the beginning of a lifelong faith-walk with our Creator and Savior.

I can use a historical theological analogy to illustrate salvation in Christ. If you are in Christ, you need not fear death and the judgment that follows because your sin debt (Romans 6:23) has been paid in full by Christ (Romans 5:8). Just as the Hebrew slaves in Egypt put the blood of a spotless lamb on their doorposts so the punishment of death would pass over them (Exodus 12:13), the blood of Christ, the sinless lamb of God, on the doorposts of your heart by faith will cause God to pass over your sins on your day of judgement.

In this work, I have tried to make a reasonable case for a creator as opposed to a totally naturalistic model of origins. I have used our common observations and knowledge of the properties of the physical universe to make my case. I have not been trying to "prove" God but merely to make a reasonable case. In the final analysis, one does not come to God by proof or wisdom but by choice. You do not love your spouse, children, parents or friends by an accumulation of facts and wisdom. In the same way, you cannot know, and love God based solely on evidence, proof or wisdom. In the Bible verses I quoted above, it is clear that God expects us to recognize him in his creation. I have merely tried to point to some of the evidence that makes such faith reasonable. However, to recognize him is not to "know" him. Like with any relationship, to truly know God you must trust him, humble yourself and seek him on his terms. Spend time with him and in his Word (the Bible). If you truly seek to know God on his terms, you will find him (Hebrews 11:6).

Anyone can become a Christian by praying to receive Christ as their personal Savior. This is the first step in faith and trust but there is much more to the Christian life and it's important to find a Bible believing church where you can find fellowship and instruction. Read and study the Bible on a regular basis. Christians are not perfect; we are not sinless, but we are forgiven by God's grace and our faith as we confess our sins and walk with our creator.

In Christianity, the strong, wealthy and the intellectual have no advantage over the weak, poor and common person before God. We all access God and receive His salvation on His terms, by faith. We are one people in Christ.

Ephesians 2:10 *"For we are his workmanship, created in Christ Jesus for good works, which God prepared beforehand, that we should walk in them."* It's important to remember that we are not "saved" by doing "good works." We are "saved" by faith to do "good works" to honor our God who is our Creator and Savior.

FOR FURTHER STUDY

CREATION WEBSITES

www.icr.org
www.answersingenesis.org
www.trueorigin.org

INTELLIGENT DESIGN WEBSITES

www.arn.org
www.discovery.org/id/

ONLINE RESOURCES

www.blueletterbible.org
www.rzim.org
www.gnpcb.org/esv/
www.e-sword.net

Evidence That Demands A Verdict: Life Changing Truth For A Skeptical World, by Josh and Sean McDowell, PhD published by Harper Collins.

The Genesis Record: A Scientific And Devotional Commentary On The Book of Beginnings, by Henry Morris PhD, published by Baker Books.

THE MIRACLE IN YOUR BACK YARD

The Monarch's Journey from Caterpillar to Chrysalis to Butterfly

— If you ate like a monarch caterpillar, your favorite food would be poisonous to most other animals, but you would rather starve to death than eat anything else.

— If you grew at the same rate as a monarch caterpillar you would weigh just over eight tons by the time you were 20 days old. That's almost four times as heavy as my Toyota pickup. Aren't you glad you don't have to pay that grocery bill?

— You can see well with two eyes, but did you know that the monarch butterfly has two compound eyes with 6,000 simple eyes in each? That's 12,000 eyes that can see all the colors we can see plus ultraviolet light, which we can't see. They can also detect polarized light. Can you imagine what they see through 12,000 eyes?

— To taste your food the way a monarch butterfly does you would have to take off your shoes and socks and stand on your plate. The monarch butterfly smells with its antennae and tastes with its feet. Aren't you glad you don't have to stand on your pizza to taste it?

– Have you ever wished you had an extra brain? The monarch butterfly has seven extra nerve centers distributed throughout its body that serve as mini-brains. These seven extra "brains" allow the monarch butterfly to live, walk around and even fly after its head has been cut off. (Don't try this at home.)

This is the story of the amazing life cycle of the monarch. The phenomenon of metamorphosis baffles evolutionists, inspires creationists, and astounds just about everybody. The life-story of the monarch stands out as much more than evidence for intelligent design – it points to an ingenious Creator with an eye for beauty. God has equipped the monarch, a "dumb" insect, with the hardware, software, and firmware necessary to undergo amazing transformations and perform astounding feats. When we take a close look at the monarch, we can't help but "see" the Creator.

Much of the information contained herein was obtained from various Internet web sites and, most importantly, from a book by Jules H. Poirier titled *From Darkness to Light to Flight – Monarch – the Miracle Butterfly*. This informative book is available on Amazon.

Where should I begin this story? The monarch life cycle begins with an egg out of which hatches a caterpillar. The caterpillar transforms into a chrysalis. The chrysalis transforms into a butterfly which lays an egg that repeats the cycle. I could begin anywhere in the cycle, but I'll start with the caterpillar. I used to see these caterpillars on milkweed plants on the ranch where I grew up. As a kid, I had no idea of the journey those monarch caterpillars were on.

THE MONARCH CATERPILLAR

Monarch caterpillars can only live on plants in the milkweed family. Their digestive tract is designed to digest only milkweed so they will actually starve rather than eat any other plant. Fortunately, there are over 220 distinct genera and 2,400 species of milkweed available for their dining pleasure. Most species of milkweed are toxic to other plant-eating animals. Since the monarch caterpillar eats only milkweed, it carries milkweed toxins in its body, as does the butterfly that follows. This makes them toxic or at least bad tasting to predators. Their toxic food and toxic taste is a good defense mechanism for both the caterpillar and butterfly. Consequently, rather than having colorations that allow them to blend into their environment so they can hide from predators, the toxic caterpillar and butterfly carry bright and beautiful markings as a warning that tells predators to, *"Stay away, I'm a toxic monarch."*

A female monarch butterfly lays between 400 to 800 eggs during her lifetime. They are laid one at a time, glued to the bottom of a milkweed leaf. Before laying an egg, the butterfly tests the plant to ensure it is suitable for her caterpillar "baby." She taps or "drums" on the leaf with her small front legs, which are equipped with sharp needle points on the end. The needles poke holes in the leaf causing plant juices to flow. The monarch then "smells" the juices with her antennae and tastes it with her feet to ensure it's suitable for egg laying. If it passes the test and she sees no predators (spiders, etc.) she will glue a single egg to the underside of a leaf.

The tiny egg contains all of the genetic information for the caterpillar that will hatch from it, the chrysalis that will follow and the butterfly that will follow after the chrysalis. It also contains the instructions necessary to direct the butterfly to fly thousands of miles to a summer or wintering location it has never seen.

The egg usually hatches in about three days and out crawls the tiny but fully functional caterpillar. The caterpillar eats almost constantly until it transitions into a chrysalis. When it initially emerges from the egg it is white, but it soon sports the colorful yellow, black and white markings and four black feelers that distinguish the adult monarch caterpillar. The caterpillar's eyes are simple, only able to distinguish light and dark, no color. The caterpillar's head has a hollow tube called a spinneret, used for passing wet silk thread coming from the silk gland inside the body. When exposed to air, the wet silk dries hard in about two seconds. The thread is strong enough to easily support more than the weight of the caterpillar.

The caterpillar has three pairs of front legs ending with claws, and five pairs of prolegs toward the rear of the body. The round prolegs are tipped with suction cups, which allow the larva to stick to the surface as it walks. The six front legs are used mainly in eating and shedding the head capsule while the prolegs are used for locomotion. The last two prolegs are called claspers, because – like hands – they can clasp things. The caterpillar also has four feelers, two at each end, that are used to sense objects.

The caterpillar's heart is located toward the rear of the body and beats about once per second. The caterpillar's green blood is pumped through a tubular heart from the rear to the front of the body. The blood passes through the central body cavity back to the rear, providing necessary nutrients along the way. This is called an open circulation system because it doesn't use blood vessels.

About three days after coming out of the egg, the caterpillar goes through the first of four or five molts, shedding its outer skin as it increases in size. Prior to molting the caterpillar is usually quiet for several hours. Then it spins a silk pad and attaches itself to it with its rear clasper prolegs. It then spins another silk pad in front of its head. It hangs on to the front silk pad with its front legs and pulls itself forward until the outer skin breaks loose. It then works its way out of its outer skin. The caterpillar uses its front legs to pull off the head capsule. This leaves the caterpillar temporarily blind. The head capsule with eyes and spinneret are discarded with each molt. After a couple hours, the new outer skin hardens. The caterpillar usually eats its old outer skin, except the head capsule that is made of a hard material (chitin). Since there are four or five molts, the caterpillar must grow four or five different head capsules, each one being a bit larger to accommodate the growing caterpillar.

On the last molt, about 20 days from the time the egg was laid, the full-grown caterpillar is about two inches long and weighs around 1.5 grams. This weight is 2,700 times greater than when it first hatched from an egg. This is equivalent to a seven-pound human baby growing to 18,900 pounds (over nine tons) in 20 days. This growth rate is unique to this class of animal.

THE MONARCH CHRYSALIS

After about 20 days of nearly non-stop eating, the caterpillar is ready to transition into a chrysalis. It stops eating and begins to hunt for a suitable site for the transition. It looks for a sturdy place that is sheltered from bright sun and rain. The site must enable the caterpillar to hang upside down without touching anything. Often it picks a location on the underside of a leaf in the interior regions of a milkweed plant. Sometimes it will even leave the milkweed plant to find a suitable location elsewhere. Once a suitable location is found, the caterpillar spins only one silk pad. It attaches to this silk pad with its two rear clasper prolegs and hangs upside down with the head slightly elevated forming a "J" shape.

After this, except for occasional movements, the caterpillar is usually very still for about 12 hours. When it's getting close to the time for the caterpillar to molt into a chrysalis it will lift its head toward the body and hold this position for several seconds – as if doing sit-ups. These "sit-ups" tend to occur with increased frequency as the transition approaches. When the time comes, the transition from caterpillar to chrysalis takes place quickly – in about a minute.

The transition begins with a split in the caterpillar's skin just behind the head. The green chrysalis that has been forming beneath the skin appears. The caterpillar begins to jerk and twist violently as the outer caterpillar skin folds up toward the silk pad, exposing more of the green chrysalis. When the caterpillar's skin is rolled up to the top a black stalk-like apparatus extends from a hole in the chrysalis' abdomen and inserts into the silk pad. This stalk-like device is called a "cremaster" and it has a bulb on the end making it look something like a small black Q-tip. The bulb end of the cremaster is covered with hundreds of tiny barb-like hooks. When the chrysalis has plunged the cremaster into the silk pad it gyrates in a circle a time or two to make sure the hooks are firmly embedded in the silk pad. It continues to gyrate until the old skin falls away leaving the chrysalis alone hanging from the cremaster stem.

Free of its caterpillar skin, the chrysalis hangs quietly for about eight days while the monarch butterfly takes shape within. It can take up to a month for the butterfly to form and emerge if the weather is cold.

The length of the chrysalis is about half the length of the caterpillar. The chrysalis has a yellowish band around the abdominal area. Within about 16 hours this yellow band will turn to around 24 colorful spots that look like metallic gold underscored by a thin black line. There are

14 other yellow spots on the outside body of the chrysalis. All of these spots will also look like beautiful metallic gold as the green chrysalis darkens.

Inside the chrysalis, most of the caterpillar has turned into a green liquid but a red heart beats beneath the golden crown pumping green blood throughout the chrysalis. The chrysalis shell is largely transparent so the green "soup" that was once a caterpillar makes the entire chrysalis appear green except for the black cremaster and the gold crown and spots. It hangs like a green jewel studded with gold.

A few days before the butterfly is ready to exit it can be seen taking shape within the chrysalis. By the time it's ready to emerge the distinctive black and orange colors of the monarch butterfly are clearly seen through the translucent chrysalis skin.

The transition from egg to caterpillar to chrysalis to butterfly can take only about 30 days in warm weather. With lower temperatures the process is slower but normally the egg hatches in about three days, the caterpillar transitions to chrysalis in about 20 days and the butterfly emerges from the chrysalis in about eight days. When the weather is cool or cold it can take up to a month for the butterfly to exit the chrysalis.

THE MONARCH BUTTERFLY

When the butterfly first emerges, its wings are small and tightly folded so it must unfold and "inflate" them by pumping fluids from its body into veins in the wings. The freshly emerged monarch's fat abdomen shrinks as fluid is pumped into the wings. In about fifteen minutes the wings are fully inflated and formed into their proper shape. The butterfly usually continues to hang for a few hours to dry, rest and adjust to its new body. In cold weather the butterfly may wait for an entire day before taking flight.

The monarch has four wings; a larger front set that spans about four inches and a smaller rear set. The top side of the wings is orange with black lines and borders and white spots along the edge. The bottom side of the wings is lighter in color.

The monarch butterfly is a typical insect with six legs and three main body components consisting of the head, thorax, and abdomen. The monarch's head includes two compound eyes, each with 6,000 individual simple eyes. Nobody knows exactly what a monarch sees with its 12,000 individual eyes. We know that monarchs are able to see all the colors humans can see plus ultraviolet which humans can't see. They can also detect polarized light, which is suspected to be an important navigation aid.

The monarch has two antennae on its head, which are used for balance in flight as well as a sense of smell. They frequently lower their antenna to smell a flower they have landed on. The male monarch can emit a scent or "perfume" which the female is able to detect with her antennae at distances up to two miles. A long proboscis or hollow tongue is used to suck nectar from flowers. Nectar is the primary food of the butterfly, which is no longer limited to the milkweed plant. When not in use the proboscis is rolled up out of the way.

Six legs and four wings are attached to the butterfly's thorax. Two of the six legs are very small and mounted just below the head. The four larger legs are used for walking. The large legs end with a claw that allows the butterfly to cling to various surfaces. These legs also include taste sensors.

The monarch butterfly's abdomen includes the heart, sex organs, and digestive system. The heart pumps the monarch's green blood through a tube toward the head where it is released to flow back through an open circulatory system to the abdomen providing nutrients to the various body parts along the way. The butterfly breathes through air tubes (spiracles) on the thorax and abdomen.

Like other butterflies, the monarch sometimes looks awkward as it flutters about. In reality, however, monarchs are expert high-tech flyers. Their unique appearance in flight is derived from the fact that they are choosing each wing stroke from a repertoire of movements that generate or take advantage of a wide variety of aerodynamic mechanisms. With wing strokes that produce extra lift from vortices or take advantage of air movement from previous strokes, the monarch can fly as fast as 30 miles per hour and have been seen as high as 12,000 feet above ground.

Monarchs have eight nerve centers distributed throughout their body. The primary nerve center is the brain but other nerve centers control wings, legs, digestion, etc. Because of these secondary "brains" a monarch can continue to fly and walk after the head has been removed.

One of the most amazing things about a monarch butterfly is its ability to migrate thousands of miles to a precise location it has never seen. There are two migrating populations on the North American continent. These populations are divided by the Rocky Mountains. In the fall both populations migrate to separate locations in Mexico, sometimes traveling as much as 3,000 miles. Some western monarchs winter in temperate locations along the California coast. After wintering in Mexico or the California coast, they return to their summer habitat in the spring. Of course, not every monarch butterfly lives long enough to make the entire circuit, more often its their children or grandchildren that complete the cycle but every generation "knows" which direction and where to go when the time comes.

Monarchs born during the hot summer mature rapidly, become sexually active and typically live only a few weeks or months. Monarchs born during the fall usually don't mature as rapidly, don't become sexually active and can live up to a year. They migrate to their wintering site and at least part way back to their summering location during the spring. Thus, monarch butterflies live from six weeks to nearly a year depending on when they are born, environmental conditions, how long they stay in their wintering location and how rapidly they mature. The migration pattern continues through multiple generations with each generation continuing the migration in the correct seasonal direction.

The monarch's unique navigation system enables them to migrate to a precise location even if a wind blows them hundreds of miles off course during their journey. Scientists are not certain of the precise navigation technology monarchs use but it's believed that they use a combination of two navigation systems. The first system is an internal compass that uses the earth's magnetic field. Monarchs have tiny particles of an organically deposited magnetic material called magnetite in their head and thorax region. This would allow them to orient their direction relative to the earth's magnetic field. Secondly, with their ability to see polarized light, monarch butterflies can navigate relative to the sun's position – even on cloudy days. In combination, these navigation systems along with genetic programming allow monarchs to navigate with great accuracy.

The Creator has equipped the monarch butterfly with all the necessary hardware, software and firmware it needs to find its way on its life journey. Today, with the aid of clocks, calendars and mathematics, humans can navigate using the position of the sun. With a compass, we can navigate using the earth's magnetic field. With airplanes we can fly. After 6,000 years of technological development we humans can fly and navigate almost as well as one of God's insects. Recognizing our propensity to stumble in the dark, drift off course and get lost, the Creator has equipped us to successfully complete our life journey. God created us in His image with a will and a mind that can reason. Then He gave us His Word, the Bible to be our light and compass to guide us through life to a place we have never been before.